THE CYCLING FEMALE

THE CYCLING FEMALE

HER MENSTRUAL RHYTHM

Allen Lein

UNIVERSITY OF CALIFORNIA, SAN DIEGO

 W. H. FREEMAN AND COMPANY
San Francisco

Sponsoring Editor: Arthur C. Bartlett
Project Editor: Nancy Flight
Copyeditor: Susan Weisberg
Designer: Marie Carluccio
Production Coordinator: Linda Jupiter
Illustration Coordinator: Batyah Janowski
Artist: Donna Salmon
Compositor: Typesetting Services of California
Printer and Binder: The Maple-Vail Book Manufacturing Group

Library of Congress Cataloging in Publication Data

Lein, Allen, 1913–
   The cycling female.

   Includes index.
   1. Menstrual cycle. 2. Menstruation disorders.
I. Title.
RG161.L44   618.1'72   78-23675
ISBN 0-7167-1039-0
ISBN 0-7167-1038-2 pbk.

Printed in the United States of America

2   3   4   5   6   7   8   9

for David and Laura —
with love and admiration

CONTENTS

PREFACE ix
CAST OF MAIN CHARACTERS xix

1 KINDS OF REPRODUCTIVE CYCLES 1

How the Birds Do It 2
Cycling in Mammals 4
A Preliminary Look at the Human Cycle 9

2 THE PARTS OF THE CYCLING SYSTEM 11

The Uterus 15
The Ovaries 18
The Pituitary Gland 22
The Brain 28

3 PUTTING IT ALL TOGETHER 41

Some General Principles 41
The First Half of the Cycle: The Estrogenic Phase 54
Ovulation 59
The Second Half of the Cycle: The Progestational Phase 64
Cycles Without Ovulation 70

4 WHY THE MENSTRUAL CYCLE
SOMETIMES STOPS 73

Pregnancy and Lactation 74
Menopause 78
Emotional Causes 79
Defects in the System 85

5 ARE WOMEN REALLY DOMINATED
BY THEIR MENSTRUAL CYCLES? 87

Premenstrual Problems 90
Painful Periods—Dysmenorrhea 93
Menopause 96
The Question 102

6 SOME FINAL THOUGHTS 105

The Age of Puberty 106
Sexuality and Contraception 107
Infertility 116
The "Social Regulatory Factor" 118
Changing Times 119

GLOSSARY 123
INDEX 133

PREFACE

In the course of planning this little book, I decided to write it mainly, although not exclusively, for young adults. In fact, as I wrote I frequently imagined that I was talking to my own teenaged children and their friends. I knew that I was going to enjoy the writing, and I hoped that it would make both pleasant and informative reading and that the inevitable technical parts would be neither overwhelming nor dull. I decided to write mainly for young adults but hoped that others might be interested also.

I must emphasize that this short work is not meant to be a textbook, although I suppose that it could provoke discussion in some classrooms. My intentions are to describe the menstrual cycle, and to emphasize its functioning and control as part of an intricate system, but without neglecting some of the emotional and social issues relating to menstruation.

You, the reader, and I might both well ask, Why devote a whole book—even a short one—to so confined a subject as the menstrual cycle? Who cares enough to want to know? Moreover, why should this particular function be discussed by a man, who obviously has no first-hand experience with this exclusively female function? All of these questions are worthwhile, and I will attempt to answer them before delving into the subject itself.

Why a book about the menstrual cycle? First, perhaps because it is so well defined; its external and internal manifestations can be detected and measured, and we have learned a great deal about its control. In addition, the menstrual cycle and the menstrual period are important aspects of the lives of all women. Many women see menstruation as a sign that their bodies are functioning properly, although some think of menstruation as a nuisance at best and some commonly refer to it as "the curse." Many men seem to feel that there is something mysterious about menstruation and tend to regard it with repugnance and perhaps some fear. Ignorance and misinformation have generated attitudes that have been and are damaging to the position and role of women in both primitive and advanced societies.

I regard the menstrual cycle to be a beautiful example of a fundamental property of nature: cyclicity. We humans exhibit many kinds of biological cycles, each with its own set of controls and with its characteristic time period. For example, our hearts beat cyclically (rhythmically), we sleep at more or less regular intervals, and some of our glands have 24-hour cycles. All living things cycle.

In fact, conception, birth, life, and death may be regarded as a kind of cycle.

Nonliving things also cycle; the earth itself exhibits well-defined cycles—day–night, the seasons. There are some cosmologists who claim that the entire universe, usually considered to be boundless and timeless, undergoes a grand cycle whose duration is measured in countless billions of years. Many biological cycles are synchronized with the day–night cycles of our planet and are known as circadian or diurnal cycles or rhythms.

The length of the menstrual cycle is measured in days and averages about 28 days. This approximates the length of the lunar month, and the word *menstruation* is derived from the Latin *mensis,* which means month.

Still another reason for focusing on the menstrual cycle is the recent discovery that it is controlled by the brain. In fact, the brain controls most reproductive function—not just sexual behavior, in which the role of the brain is obvious, but the onset of sexual maturity, or puberty; the timing of ovulation and the menstrual period; milk production; and so on. Recognition of this role of the brain represents one of the most fascinating breakthroughs in modern physiology; it has fundamental implications, and it opens the door just a crack to possibilities of control that both involve and boggle the mind.

Let us now rephrase the question concerning who cares about the menstrual cycle by asking, Who should care? The answer is: We all should. After all, about half the human population is women, virtually all of whom will have menstrual cycles over a period of roughly 35 years. It seems only logical that women should want to know—should want to understand—this intriguing function that has been taboo as a topic for discussion for much too long. How about men? Should they want to know? Of course. Even those of us who like to flatter ourselves that we "know about women" cannot pretend to have even modest understanding unless we know something about the menstrual cycle, a condition and process that for humans is uniquely the province of women.

Isn't this subject, then, better handled by a woman? I would have to agree that the more personal and subjective aspects of menstruation can probably be discussed only by women; after all, no man knows exactly how it feels to have a menstrual cycle. However, the subject matter to be considered in the following pages is largely, although not exclusively, descriptive of the objective, observable or measurable aspects of the menstrual cycle and its control, and these sections will therefore rarely reflect the gender of the author. In further self-defense, let me add that I have spent well over half of an already long life in the study of those glands that are involved in regulating functions such as the menstrual cycle, and I take this means of sharing some of the knowledge, opinions, and attitudes I have accumulated over the years. In recent years several books that include discussions of reproductive function and sexuality in women have been written by women, and I am pleased to refer to some of them:

> *The Curse* by Janice Delaney, Mary Jane Lupton, and Emily Toth (New York: Dutton, 1976).
> *Menstruation and Menopause* by Paula Weideger (New York: Knopf, 1976).

Our Bodies, Our Selves by the Boston Women's Health Book Collective, 2d ed. (New York: Simon and Schuster, 1976).

Let me emphasize that this book is not intended to be a "how-to" manual. It will not instruct in sexual techniques, although from time to time I will discuss sexual activity as it pertains to the cycle. I will not offer medical advice or recommend contraceptive methods, but menstrual problems and birth control both will be discussed. My aim here is to present modern, up-to-date information on the human menstrual cycle, how it is controlled, and its impact on the day-to-day lives of women. I hope that for the readers of this book, females and males of all ages, an improved understanding of human female reproductive function will result in an enhanced ability to regard the menstrual cycle for what it is—an exquisitely controlled system for the monthly release of an egg and renewal of the "nest" in which it will be nurtured to form a new human being. I hope also to pull this subject out of the shadows and to make it easier for young men and women to discuss the menstrual cycle and related matters with their physicians and other health care specialists, with their teachers and parents, and, perhaps most important, with each other.

The book may seem a bit technical from time to time. Some of the features of the menstrual system are indeed somewhat complicated, but I believe that they are quite readily understandable. In Chapter 3 I have attempted to assemble the parts of the menstrual system (discussed in Chapter 2) and to show you how these elements of the system work together to create the cycle. You will find that throughout I have used the terms commonly used by all health professionals for the structures and processes involved in the cycle. Many, perhaps most, of them are likely to be familiar to you already.

I trust that the foregoing paragraphs help to introduce this subject that I find so intriguing. If I have omitted any questions that come to mind, I ask your indulgence and hope that answers may appear on the pages to follow.

acknowledgments

This book was begun in a tiny, secluded study perched on a cliff overlooking Lake Como, near the village of Bellagio in northern Italy. The study, known as Studio Monserrato, is situated over a small chapel on the grounds of the Villa Serbelloni, owned and operated by the Rockefeller Foundation as a site of international conferences and for visiting scholars who wish isolation and uninterrupted time for writing. These benefits were certainly provided, as were a place of unmatched beauty, a superb cuisine, and stimulating dinner company. I am grateful to the Rockefeller Foundation for creating an environment in which writing was a pleasant and almost effortless pastime.

My thanks go also to Dr. Samuel S. C. Yen for the encouragement and stimulus he offered. I have taken advantage of the patience and knowledge of many friends and colleagues who have read portions of this book and provided many valuable comments and suggestions. As the book approached completion, a few "volunteers" undertook a critical review of the entire manuscript, and I am pleased to express my appreciation to Ms. Judith A. Staples, Ms. Nancy Levy, Dr. Robert Resnik, and Dr. Doris A. Howell for having incorporated this additional task into an already busy schedule. I am grateful also to Ms. Deborah Eck for her secretarial assistance and for the preparation of the final typescript. Finally, I am particularly happy to thank my wife, Teresa, who has read the manuscript many times and has served as counselor, critic, stenographer, and enthusiast.

November 1978   *Allen Lein*

CAST
OF MAIN
CHARACTERS
(in order of appearance)

These are some of the chemical signals (hormones) the body uses in creating a menstrual cycle and to which frequent reference is therefore made. These are the names commonly employed by scientists and physicians, and they are listed here for your convenience.

HORMONE	PRIMARY SOURCE
Estradiol	Ovary
Progesterone	Ovary
FSH—Follicle-Stimulating Hormone	Anterior Pituitary
LH—Luteinizing Hormone	Anterior Pituitary
PRL—Prolactin	Anterior Pituitary
LRF—Luteinizing-Releasing Factor	Hypothalamus

THE CYCLING FEMALE

KINDS OF REPRODUCTIVE CYCLES

Not all living things reproduce sexually, but most of those that do have some sort of reproductive cycle. Sexual reproduction obviously means that both a male and a female are involved and that, by some means, male cells (spermatozoa) are made available to fertilize a female cell (egg or ovum). I shall confine our discussion to cycles in the female despite the fact that in many species, including the human, some kind of male cycle may also be found.

how the birds do it . . . .

In some forms of animal life—for example, in many birds, reptiles, amphibians, and mammals—a seasonal periodicity marks the reproductive cycle. Birds are particularly interesting because their migratory patterns, as well as their reproductive cycles, may be tied to the seasons. In most cases the springtime represents the breeding season. Before that time, in the winter, the female sex glands—the ovaries—may be small and relatively functionless; as spring approaches, the ovaries are activated and become functional. Finally, the female bird accepts a male, receives a store of sperm cells, and begins to ovulate, which means that she starts to lay eggs.

A fascinating sidelight in the case of avian (bird) reproduction is the fact that many wild birds lay a characteristic number of eggs typical of their particular species. For example, a given kind of wild bird may lay three eggs, usually no more or less. An obvious question arises: How does the bird know when she has laid the number of eggs appropriate to her species? Does she know when the correct number has been deposited, or is it that she has only that number to lay and no more? Answers to these questions were obtained through some rather devilish experiments. One scientist, after observing that a nesting bird had laid its quota of eggs, stealthily stole one from the nest. The bird responded by laying another egg to replace the one that had been removed. Each time an egg was taken, the bird was able to produce a replacement. How do such birds know when enough is enough? One is tempted to propose that they are able to count, but the fact is that they are able to tell mainly by the feel of the nest as they sit on their eggs. This ability to keep replacing stolen eggs is not inherent in all birds; some simply lay their quota and then stop, regardless of whether or not all the eggs remain in the nest.

I find that I have digressed a bit. Let us return to the seasonal reproductive cycle. How does the bird recognize that springtime is approaching and the time for breeding is near? What signal triggers the activation of the ovaries in the females and the testes in males (who, by the way, also exhibit seasonal sexual periodicity)? In some, but by no means all birds, the increasing length of the days and the resulting increase in the exposure to daylight are the signals. Early sexual activation can be induced in such birds by exposure to longer hours of artificial light at a time when their sex organs would normally be quiescent.

Even domestic hens are not immune to this kind of stimulus. Chicken and egg farmers discovered that leaving a light on in the hen house would keep the hens on a schedule of producing eggs at a time when production might otherwise decrease. There is a rather amusing story about one investigator who believed that light was only indirectly responsible and worked primarily because the hens did not sleep as much and were active for many more hours, ate more, and, as a consequence, laid more eggs. So he performed an experiment that involved a fiendish kind of machine consisting of a stick on an endless belt driven slowly by an electric motor. With the lights turned off, the stick would pass across the roost periodically, brushing the hens off and, of course, disturbing their usual hours of rest. As you might well imagine, during the first night the machine was turned on, there was great consternation in the hen house, marked by much flapping of wings and irate cackling. But after a few nights, peace appeared to have returned, and all was quiet. The clever scientist went out to investigate and found that he had been outsmarted by his hens, who had discovered the rhythm of the stick and were able to lift first one leg and then the other while the motor-driven stick passed harmlessly beneath them. Not only did the hens outsmart him, but the scientist's hypothesis was wrong as well; light does have a direct effect on seasonal periodicity in many species of birds.

cycling in mammals . . . .

Light also plays a role in the sex cycle of some mammals, and perhaps we should discuss mammalian reproductive cycles in general before focusing specifically on the human. Many mammals have seasonal cycles similar to those of birds; they are sexually quiescent during much of the year but become sexually active at the time of the breeding season. Other mammals are sexually active and fertile the year around, and humans, of course, are among these. In general, two kinds of cycles are found among mammals: those in which ovulation (release of ova) occurs spontaneously (as in the human) and those in which ovulation normally occurs only after externally applied stimulation. Let us consider the latter type first.

Rabbits and cats are noteworthy examples of animals that prepare to ovulate several times each year. They may remain prepared over some extended period and are sexually receptive during this time. The act of copulation normally serves as the stimulus to ovulation, and these animals will ovulate a few hours following the sex act. This system is highly efficient from the point of view of reproduction since it virtually insures the availability of mature ova at the time the spermatazoa are deposited, and thus it yields a high pregnancy rate.

A question of great interest is how the ovary receives and recognizes the signal that results in ovulation. The initial stimulus involves sensitive sensory nerves in and around the genital region. In fact, the simple insertion of a glass rod or similar object into the rabbit's vagina will usually trigger ovulation. Sometimes simply body contact with or the mere presence of a male rabbit will induce ovulation. The pathway for the chain of events between the initial stimulus and the ultimate ovulation involves the brain and will be discussed in a subsequent chapter.

As I mentioned earlier, the system just described does not operate in humans. I know of no reliable evidence to indicate that sexual activity of any kind (anything from preliminary sex play to actual intercourse) is able to induce ovulation in the human female. Ovulation normally occurs in women whether or not they are sexually active. A question frequently asked is whether orgasm in the female is required for or even facilitates ovulation. The answer is *no*. Fertility obviously requires ovulation, and women who do not achieve orgasm are as fertile (ovulate as regularly) as those who do. In other words, ovulation in humans is not correlated with either the qualitative or quantitative aspects of the female sexual response.*

Let us now continue by taking up in some detail an example of spontaneous ovulation. The most detailed study of the female sex cycle has been made with the laboratory rat. The rats employed for such studies are highly inbred and are raised specifically for use as experimental animals. Even more important is the fact that, although the rat does not have a menstrual cycle or a periodic loss of blood typical of the menstrual period, it does have a very well-defined cycle, known as an estrous cycle, which possesses some particularly interesting features. The cycle in the laboratory rat is quite short, either four days or five days. This does not mean that it will vary from four to five days for any given rat; rather, for a given breed of rats kept under controlled conditions, the period of the cycle will be one or the other and will remain constant at that particular interval.

On only one of the days of the cycle is the female rat sexually receptive and willing to copulate. This day is known as the day of estrus** and ovulation is synchronized with that day. Here again there is a significant difference in the character of the rat's cycle from that of

*Circumstantial evidence has led some scientists to conclude that ovulation in the human female may sometimes be induced. Although this possibility deserves further study, most of the data currently available indicate that human ovulation is ordinarily spontaneous and occurs at a particular time in the menstrual cycle.

**Animals in estrus are also said to be "in heat"; the latter term is commonly used for farm animals and pets.

the human. Whereas the female rat will copulate *only* on her day of estrus, women are sexually active at any time during their cycle. Some claims have been made that the intensity of the sex drive (libido) in women may vary in synchrony with the menstrual cycle, but the data appear to be inconsistent. Some reports indicate that sex drive is highest at mid-cycle at the time of ovulation; others state that it occurs just before and during the menstrual period. Still other investigators report no significant fluctuations in libido during the course of a menstrual cycle. It is not at all unlikely that libido in women is influenced more by a hormone usually considered to be a male hormone (testosterone) than by those hormones involved in controlling the menstrual cycle. (Testosterone is produced by the male sex glands, the testes, but is also produced in significant amounts in women by their adrenal glands and even by their ovaries.) There is an obvious and substantial interest in the possibility of cyclicity in female sex drive, but the methods that have been used for measuring libido in women may be of questionable reliability and validity.*

As a parenthetic note, I should mention that, although many men and women tend to avoid sexual intercourse during the menstrual period, presumably for esthetic reasons, there is no physiological reason for doing so. Those who are not offended by menstrual blood find that sex can be quite satisfying at this time; moreover, the appropriate use of a diaphragm can temporarily minimize or eliminate the blood ordinarily found in the vagina during the menstrual period.

*A *reliable* method is one in which repeated measurements give the same answer within reasonable limits. A *valid* method is one that measures what it is intended to measure.

Let us now return to the rat, because for this usually abhorred but throughly investigated little animal there is an environmental factor that plays a particularly important role in the timing of its ovulation. That factor (remember the hen house) is light. Suppose that a group of "four-day" laboratory rats is kept under controlled illumination. The lights are turned on at 6:00 A.M. and off at 6:00 P.M.; accordingly, these rats have 12 hours of light alternating with 12 hours of darkness. Not only will these female rats have a day of estrus every four days, they will always ovulate in the very early morning of that day, approximately between 2:00 and 4:00 A.M. The stimulus for that ovulation occurs about 12 hours earlier, in the afternoon of the previous day. If the rats are put to sleep with an anesthetic for a few hours during that critical afternoon, ovulation fails to occur the next morning and is postponed almost exactly 24 hours. Here, then, is a reproductive cycle tightly coupled to the daily, or diurnal, cycling of day and night.

What would these rats do if their days and nights were suddenly reversed and the lights turned on at 6:00 P.M. and off at 6:00 A.M? That experiment has been performed many times and has shown that when the reversal of day and night is instituted, the estrous cycle may be a bit erratic for a while, but then it rapidly accommodates to the new schedule of illumination so that ovulation again occurs in the hours of darkness—somewhere around 2:00 to 4:00 A.M. "rat time," but between 2:00 and 4:00 P.M. actual clock time. If the rats are kept in either continuous light or continuous darkness, they stop cycling altogether and do not ovulate; in constant light, they go into continuous estrus, while in constant darkness they ultimately fail to go into estrus at all.

The important environmental signal in the normal control of the estrous cycle in the rat is clearly the diurnal light-dark cycle. The rat is not unique in this respect; remember that some birds and other animals have reproductive cycles that are light-dark sensitive. The control of the human menstrual cycle is not dependent on light, although some women report that their cycles may be a few days longer in the winter months when the days are short than during the summer months. However, for the most part the menstrual cycle in women is independent of changes in environmental illumination. Blind women may have normal menstrual cycles; blind rats, on the other hand, are noncycling—the same as those rats kept in constant darkness.

a preliminary look at the human cycle . . . .

There are, then, some similarities between the estrous cycle in the rat and the menstrual cycle of women, but there also are many differences. It should be clear at this point that both are examples of spontaneous ovulation, and in both ovulation is synchronized with the respective cycles. There is, of course, a great difference in the length of the cycles; the length of the menstrual cycle in women averages about 28 days, although this will vary from one woman to another and from one time to another within a given woman.

Menstrual cycles occur only in primates, including monkeys, apes, and baboons, but it is in humans that the menstrual cycle has been most extensively studied. The menstrual period is one part of the menstrual cycle and is characterized by a loss of blood and tissue, both having their origin within the uterus and discharged to the outside via the vagina. The process involves a monthly destruction and renewal of the inner lining of the uterus, known as the endometrium.*

Although nonhuman primates have cycles that are classified as menstrual cycles, their menstrual periods do not characteristically involve as much loss of blood and tissue as is the case for the human. Furthermore, they have fewer menstrual cycles and menstrual periods than humans, because for a substantial fraction of their reproductive lives they are either pregnant or nursing their young. (The effect of both pregnancy and lactation on the menstrual cycle will be discussed in Chapter 4.) The rat also renews its endometrium cyclically, but the old endometrium is absorbed and not expelled as it is in the human.

So much for a preliminary discussion of reproductive cycles. In the following chapter we will take up some of the principles of biological control and review the various components that make up the cycling system in women.

---

*Endometrium arises from both Latin (*endo*) and Greek (*meter* or *metra*) and means within the uterus or within the mother.

2

THE PARTS OF
THE CYCLING
SYSTEM

Before we discuss the various components of the reproductive system, it might be well to take up a more general but important and relevant issue. All reproductive
cycles, and certainly the menstrual cycle, involve the par-
ticipation of several different organs in the body. This is
not a situation unique to reproductive cycles; no body

function normally occurs in isolation without controlling or regulating influences from anatomically separate or distant organs or tissues. All of our regulated functions, such as respiration, blood circulation, reproduction, and so on, require the integrated or coordinated action of many different parts of the body.

As soon as we recognize this situation, we are faced with a fundamental question: By what means is one part of the body able to communicate with another in order to achieve a smoothly integrated function? This question has been the subject of investigation by many modern biomedical scientists, particularly physiologists. Knowledge of the details of internal communication has turned out to be very important to the modern practice of medicine since a breakdown in the communication system is responsible for many of the diseases that befall human beings. As we will see later, disturbances and irregularities in the menstrual cycle are usually caused by communication defects. The body has only two general methods of accomplishing this all-important communication. One is by means of the nervous system (neural communication); the other relies on chemical signals transported by the blood (humoral communication).

The nervous system operates something like a huge and complex telephone network with telephone wires (nerve fibers) serving as discrete pathways for transmission of signals. We also have a central exchange, the brain, unbelievably complex and able to sort out signals, receive them from, and send them to various parts of the body. In addition, we have, in general, two sets of nerves—one (sensory nerves) to send signals to the central exchange, the brain and spinal cord, and the other (motor nerves) to transmit signals from the brain to muscles and other structures. In the case of a telephone system, we activate the telephone line by taking the phone off the hook, and we send a signal to the central exchange by dialing. The central exchange then activates the line we are calling and alerts the party on that end, and the telephone channel of communication is then open and operating.

In many, but not all, cases the nervous system operates in the same way. If my hand should inadvertently contact a very hot object, sensory nerves in the hand are stimulated; they send signals to the spinal cord and the brain, which in turn activate the appropriate motor nerves going to muscles in my arm, and a rapid muscle response and withdrawal from the painfully hot object occurs. The point is that the neural system of control uses well-defined pathways designed exclusively for the purpose of communication.

The other method of communication mentioned above is sometimes called a humoral method. (The word *humor* has two definitions—one has to do with mood or temperament and the other with body fluids. The latter meaning, of course, is the one applicable to humoral mechanisms.) A number of endocrine glands* are part of the humoral communication system, and some of them are intimately involved in the menstrual cycle and its control. All endocrine glands produce chemical compounds, which are released directly into and are carried by the bloodstream (humoral transport, so to speak). These

---

*Among the endocrine glands are the pituitary in the head, the thyroid in the neck, the adrenals and ovaries in the abdomen, and many others.

compounds, called hormones, are chemical signals to which distant tissues or organs are able to respond. Whereas the nervous system has characteristics in common with a telephone network, the endocrine glands perform in a manner somewhat analagous to radio transmission. A radio transmitter may blanket an entire region with its signal, but a response occurs only if a radio receiver is turned on and tuned to the proper frequency. Endocrine glands operate somewhat similarly: They secrete their respective hormones directly into the blood, which then carries these hormones to every part of the body. Most hormones thus go everywhere the blood takes them. What, then, determines whether or not a response occurs? What is analogous to the plugged in, turned on, properly tuned radio receiver? Nature has provided a clever and effective mechanism. Those structures in the body that are designed to respond to a given hormone contain areas in or on their cells that have a great affinity for the hormone and extract it from the blood. Not only do these areas extract and bind a particular hormone, but, as a part of the response mechanism, they usually interact chemically with that hormone. So the radio receiver in biological systems is a tissue whose cells possess active receptor sites for a particular hormone or hormones.

The two mechanisms of communication, neural and humoral, sometimes operate independently, but frequently they operate jointly. In the case of reproductive cycles, including the menstrual cycle, a large part of the control is humoral, but an important component is also neural, with, as you will see, a significant involvement of some parts of the brain. Therefore, I refer to the menstrual control system as a *neurohumoral* system. Let us now examine the component parts of that system.

The essential components of the cycling reproductive system in women are the uterus, the ovaries, the pituitary gland, and a particular region of the brain known as the hypothalamus. We will take up these components in the order presented above and say something about the role that each plays in the menstrual cycle and how each is regulated.

the uterus. . . . .

As you read about the uterus, please refer to Figure 2.1, which is a diagram of the pelvic area of a woman, viewed as though the body had been divided into two halves right down the middle and we are looking at the left half. This kind of portrayal is commonly used in anatomy books and has the advantage of showing the spatial relationships among the various pelvic organs.

The uterus, or womb, is roughly pear-shaped and about three inches long, with its smaller end pointed downward. The larger upper end is pitched forward and rests over the upper surface of the bladder. The narrow opening at the bottom of the uterus is surrounded by a heavy muscle, and that entire area, known as the cervix, protrudes slightly into the vagina. The vagina, an elastic tube leading from the cervix to outside the body, is that part of the female reproductive tract that receives the penis of the male during sexual intercourse and is the canal through which a newborn infant is delivered.

The cervix is almost always examined when a woman sees her gynecologist. The gynecologist can actually expose the cervix to view by gently distending the

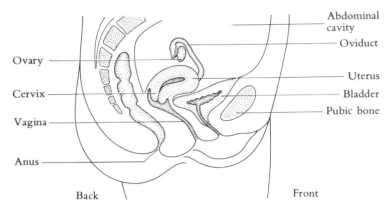

FIGURE 2.1

The pelvic organs of the female. This is a view made as though the body were cut down the middle. Note that the uterus is tipped forward over the upper surface of the bladder. The lower end of the uterus, the cervix, protrudes into the vagina. Compared to the soft and pliable walls of the vagina, the cervix feels like a rather hard knob with a depression in the middle. The oviduct is shown coming off the far side of the uterus and terminating near the ovary on that side.

vagina with an instrument known as a speculum and can also feel it, as well as the body of the uterus, by carefully holding it between gloved fingers gently inserted into the vagina, with the other hand pressing on the lower abdomen. The cervix feels like a relatively hard knob with a small depression in the middle. The gynecologist will usually wipe a few cells off the cervix and the wall of the vagina with a small swab so that these cells can be examined under a microscope. This procedure, known as the Pap smear, was introduced in 1943 by Dr. George Papanicolaou. The Pap smear is useful for determining the presence of abnormal cells, which may suggest or indicate the presence of cancer, and it is usually recommended that every woman over the age of 18 have a Pap smear every year or two.

The body of the uterus consists of two principal layers—an outer layer, composed of several strata of muscle fibers, some running up and down the length of the uterus and others circling around the uterus; and an inner layer, which lines the cavity of the uterus. Most body cavities that directly or indirectly open to the outside—the mouth and nose, for example—have a tissue lining them known as a mucous membrane or a mucosa. The uterus is no exception; its inner lining is also an example of a mucous membrane and is known as the uterine mucosa or, more elegantly, the endometrium. The vagina also has a similar lining (the vaginal mucosa), and both it and the endometrium go through cyclic changes coordinated with the menstrual cycle. The cyclic changes in the endometrium are particularly significant because the monthly loss of that endometrium marks the menstrual period.

The muscles of the uterus, known as the myometrium, are of particular importance at the time of birth since they are involved in expelling the baby from within the uterus. However, these muscles also undergo spontaneous contractions even in a nonpregnant woman. The pattern of these contractions is different for different parts of the menstrual cycle.

During the second half of the menstrual cycle, before the beginning of a menstrual period, the endometrium is thick and spongy and contains a large number of mucus-producing glands. It is then able to receive and hold the fertilized egg (which goes through several cell divisions before it reaches the uterus). The endometrial reception of the already developing embryo is known as implantation or nidation.* However, if the ovum or egg has not been fertilized and implantation has not taken place, the endometrium begins to break down and to strip away from the inner wall of the uterus, leaving a very thin mucosal layer behind. Thus begins the menstrual period. You can see why a famous embryologist has referred to menstruation as "the weeping of a frustrated uterus." As the endometrium breaks down, large numbers of blood vessels are broken, and considerable bleeding consequently occurs. With each menstrual period one to three ounces of blood are lost, sometimes more. When blood loss is excessive, five or six ounces or more every month, there is a possibility that anemia may develop, and supplemental iron may be required. In addition to blood, the menstrual flow contains considerable mucus—the product of the large number of endometrial glands—and bits and pieces of the endometrium itself.

the ovaries . . . .

The ovaries, located on each side of the uterus in the lower abdominal cavity (see Figure 2.1), have two main functions: the production and release of mature eggs (ova) and the synthesis and secretion of at least two hormones. Both of these functions occur in cycles, and it is

*Quite appropriately, nidation comes from the Latin word *nidus*, which means nest.

common to refer to an ovarian cycle, which is closely synchronized with the menstrual cycle.

The ovaries of a newborn girl may contain about 500,000 immature ova. Strangely enough, that number continuously decreases, so at the time the girl begins menstruating, perhaps at age 12 or 13, she may have only about 60,000 or 70,000 ova remaining. This is a number far greater than will ever reach maturity and become fertilizable. Assuming that a woman ovulates every 28 days on the average, and that most of the time only one ovum is released, she will "use up" about 13 ova each year. Assuming further that women ovulate over a period of about 35 years, each woman actually brings to maturity and releases only about 455 ova in her lifetime.

This is a far more conservative and economical system than is found in the male. With each completed sex act the male may release about 200 million sperm cells. If this were to occur twice a week over a period of, say, 40 years, the man would have produced and discharged over 800 billion sperm cells in that period.

For each menstrual cycle, several ova in each ovary may begin to mature, but ordinarily (about 99 percent of the time) a single ovum in one of the two ovaries will reach full maturity. No one knows by what means a particular ovum is selected for completing the process of maturation; however, that one ovum each month has a special destiny. The final maturation process involves the formation of a blister-like structure near the surface of the ovary known as an ovarian follicle; it is filled with fluid (follicular fluid) and contains the maturing ovum (Figure 2.2). During this time the growing follicle secretes increasing amounts of one of the ovarian hormones, estradiol,

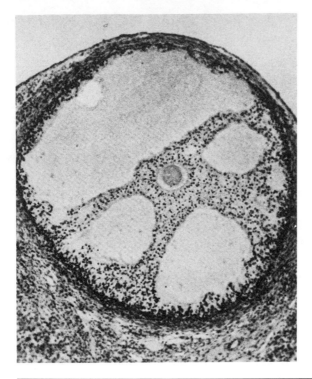

FIGURE 2.2

A slice through a mature ovarian follicle almost ready for ovulation. The ovum is the little gray circle in the center of the follicle. Most of the follicle is filled with fluid, which will escape with the ovum and some of the surrounding cells at the time of ovulation. Compare Figure 2.3. Source: From Richard J. Blandau, "Growth of the Ovarian Follicle and Ovulation." In Sturgis and Taymor (eds.), *Progress in Gynecology,* Vol. 5 (New York: Grune and Stratton, 1970), pp. 58–76. Reproduced here by permission of the authors and the publisher.

sometimes referred to as an estrogen or an estrogenic hormone. Ultimately, the ovarian follicle bursts (Figure 2.3), releasing the now mature ovum and the follicular fluid into the abdominal cavity, from where the ovum

21

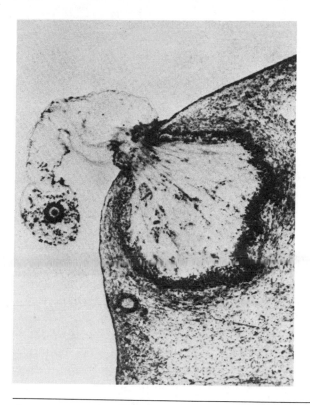

FIGURE 2.3

A mature ovarian follicle in the act of ovulation.
The wall of the follicle has ruptured, releasing
the ovum, which can be seen outside the follicle
surrounded by other cells and follicular fluid.
Source: From Richard J. Blandau, "Growth of
the Ovarian Follicle and Ovulation." In Sturgis
and Taymor (eds.), *Progress in Gynecology,* Vol.
5 (New York: Grune and Stratton, 1970), pp.
58–76. Reproduced here by permission of the
authors and the publisher.

then finds its way into a duct which leads to the uterus.

Frequently when the follicular wall ruptures, it may
bleed a little, and some blood may thus escape into the
abdominal cavity along with the ovum and follicular
fluid. Blood is very irritating to the lining of the abdomen

and can consequently create abdominal pain. Some women know when they ovulate by the pain they experience about mid-cycle. This pain is given a name taken from the German—*mittelschmerz,* which means midpain. On rare occasions the follicular bleeding may be sufficient to cause extreme pain, with a set of symptoms similar to those of acute appendicitis. It may take an astute and alert physician to distinguish between the two situations and to take the appropriate action. Some women report that this pain feels different from that which sometimes accompanies the menstrual period; the latter is frequently called a menstrual cramp and will be discussed in Chapter 5.

The ovarian follicle that has just lost its ovum now continues to develop; some of the cells within the follicle divide rapidly, filling it to form a yellowish spherical structure known as a corpus luteum (yellow body), which continues the production of estradiol and also produces another hormone, progesterone. The function of the uterus and the cyclic changes in the endometrium are dependent on the presence of these two hormones, which will be discussed in the next chapter. The menstrual cycle, including the menstrual period, results from changes in the availability of these two hormones, and they in turn are completely dependent on the pituitary gland, to be discussed next.

the pituitary gland . . . .

The pituitary gland (also known as the hypophysis) is located in a highly protected spot in the center of the head, in a little pit in the floor of the skull just below the base of

the brain. Its proximity to the brain is of particular significance because the brain exerts stringent control over
this gland. The pituitary is made up of several parts, each
with its particular function, but the two main parts are
the anterior pituitary (adenohypophysis) and the posterior
pituitary (neurohypophysis). We will be concerned
mainly with the anterior pituitary, although occasional
reference to the posterior pituitary may be made.

Over the past 50 years the number of hormones ascribed to the anterior pituitary has varied, but it now
seems that this small gland produces six different hormones, each of which has a well-defined and important
function. Three of those six hormones are involved in reproductive function, and two of those three exert profound control over the menstrual cycle.

A quick summary of the hormones of the anterior
pituitary may be useful in your understanding of the importance of this one little gland to our overall well-being.
One of its hormones, somatotropin, or growth hormone,
is responsible for normal growth; without it, the individual may mature but will remain a dwarf. On the other
hand, if the pituitary produces too much growth hormone from childhood on, a condition know as giantism
may result. Both of these situations are fortunately rather
rare, but a few pituitary dwarfs and pituitary giants are
always around.

Another hormone produced by the anterior pituitary
is adrenocorticotropic hormone, more conveniently
known as ACTH. As its name suggests, this hormone is
responsible for maintaining normal function of the outer
part of the adrenal gland, the adrenal cortex, which,
among other things, produces cortisone. Without ACTH
the adrenal cortex becomes almost functionless, and indi-

viduals with this condition become very ill and do not live long unless they are properly treated.

Still another anterior pituitary hormone is thyroid-stimulating hormone, also known as thyrotropin or TSH. Its name tells you what it does. The thyroid gland, situated at the base of the neck, relies on TSH for its normal function. Absence of TSH is rather rare; deficient thyroid function, not so rare, is usually caused by something other than inadequate TSH. Thyroid hormone is required for normal growth and development, for normal metabolism in general, and for normal brain function. With the complete absence of thyroid function from childhood on, normal growth and intellectual development does not occur, and even in the adult a total loss of thyroid hormone can lead to feeblemindedness.

Now we get to those anterior pituitary hormones that are involved in reproduction. First, let us consider a hormone called prolactin (PRL), or lactogenic hormone. Although this hormone has some influence over ovarian function, as you will see a bit later, its main action is to stimulate the properly prepared breast to produce milk. Milk production is not a part of the menstrual cycle, but it might be worthwhile to digress briefly and to discuss the function of the mammary glands, the breasts.

The development of the breasts is dependent on two ovarian hormones already mentioned, estradiol and progesterone. As girls approach puberty, the time when they become sexually mature, their ovaries begin to produce increasing amounts of these two hormones, and growth of the breasts is one of the results. This is one of the early signs of approaching sexual maturation and may be regarded with awe and delight by the young woman. Some parents may have mixed feelings; although they may take pride and satisfaction in watching the normal maturation of their daughters into womanhood, they may be reluctant to lose the little girl, a loss which, in a sense, begins with the breast development.

For some girls beginning breast development may tend to provoke a bit of anxiety. Unfortunately, in some segments of our culture, femininity is equated with the size of the breasts. Let it be understood, however, that the function of the breasts as the source of nutrients for the infant and the role of the breasts in a sexual relationship are not dependent on size. Small breasts can produce more than adequate amounts of milk and are as erotically sensitive as are larger breasts. Concerns about body development are not confined to girls; boys have equivalent anxieties about their bodies as they mature. However, for both sexes it is infinitely more important to develop into vital and fulfilled adults than to meet some stereotyped body form.

Now, let's get back to the roles of the two ovarian hormones in the development of the breasts. Estradiol is responsible for the development of the nipples, the pigmented area around the nipples, and the ducts within the breast that bring milk out to the nipples. Progesterone, on the other hand, appears to be principally involved in the development of the glandular part of the breasts, where milk is actually formed. Prolactin is able to stimulate the breasts to produce milk only if they have first been exposed to adequate amounts of estradiol and progesterone, and this is what was meant earlier by "properly prepared breasts."

Since estradiol and progesterone levels in the blood go through cyclic changes in synchrony with the menstrual cycle, it is not suprising to find that the breasts also may exhibit cyclic changes. The size of the breasts may change somewhat during the menstrual cycle, becoming a bit larger during the week or so just prior to a menstrual period. At the same time the breasts may become rather tender or even painful, or particularly sensitive to pressure. This is apparently caused by an accumulation of fluid, which results from a combination of estradiol and progesterone, both characteristically at a high level in the blood at this time of the menstrual cycle. In part because of this tenderness, self-examination of the breast is usually more successful after a menstrual period than just before the period.

It might be interesting to spend just a little more time on prolactin. It has been called a maverick among the anterior pituitary hormones because it is the only one that seems to have many different functions* and that is actively secreted without a required stimulus from the hypothalamus. In some fish prolactin is involved in salt and water balance and may play a role in permitting such fish to move between salt water and fresh water. In birds prolactin has a profound effect on behavior; it induces so-called broodiness or nesting activity. Whether or not prolactin may be invloved in maternal behavior patterns in mammals, particularly in humans, is still not clearly understood.

Let's get back to our main subject by discussing the remaining two hormones produced by the anterior pituitary gland. These have been deliberately saved until last because they are very important in the control of the menstrual cycle. These two hormones are known collec-

---

*About 80 different actions are now attributed to prolactin.

tively as the gonadotropic hormones, which means "acting on the gonads"; the gonads are the primary sex glands—the ovaries in the female and the testes in the male.

The two gonadotropic hormones are follicle-stimulating hormone, more conveniently know as FSH, and luteinizing hormone, known as LH. Both of these hormones are present in men as well as in women, but we will discuss their function in women only. The names of these hormones come close to describing their role in women

FSH stimulates the development and maturation of ovarian follicles, each of which, as you will recall from our discussion of the ovary, contains one ovum. Usually, although certainly not always, only one follicle develops to maturity during each cycle. As the follicle develops, FHS plays still another role in that it causes the ovarian follicle to produce and secrete one of the ovarian hormones—estradiol.

FSH alone can bring the follicle to full maturity, but it is not able to proceed from that point. Now LH is required. LH causes the mature follicle to rupture, releasing the ovum and all its follicular fluid, and frequently a little blood. The ovum, first released into the abdominal cavity, finds its way into a tube known as either the oviduct or the Fallopian tube (named after a sixteenth-century Italian anatomist, Gabriel Fallopius), where it is fertilized if spermatozoa are present. From this tube the ovum is conducted into the uterus. LH has by no means finished its job when it induces ovulation. The remains of the ruptured follicle go on to form a new structure, which was discussed briefly in the section on the ovary. This new structure, called a corpus luteum, produces both estradiol and progesterone. All of this activity requires the continued availability of LH from the anterior pituitary, so you see that LH is a hormone of considerable importance. The formation of the corpus luteum is said to involve a process of luteinization; thus the name luteinizing hormone.

At the risk of introducing some confusion, I should mention at this point that in some animals LH causes the formation of the corpus luteum but is not able to make it produce its hormones. The laboratory rat is an example. In this case the maverick pituitary hormone prolactin takes over and stimulates the corpus luteum to produce its hormones. It is not clear whether prolactin has a similar function in humans, but possibly it, as well as LH, is able to keep the corpus luteum active in producing estradiol and progesterone. At any rate, the ovaries are completely subservient to the anterior pituitary gland. Without a functioning anterior pituitary, the ovaries also are functionless, producing neither hormones nor mature ova.

the brain . . . .

It is sometimes said that we have two brains, and in a functional sense it is possible to divide the brain into two components. However, this is a somewhat arbitrary division, and the two parts interact very closely with one another.

One part of the brain, by far the largest, is the cerebrum, which itself is divided bilaterally into two halves. This uppermost part of the brain is the seat of consciousness, perception, and intellectual activity and is the part of the brain most recently acquired in the course of human evolution.

Under the cerebrum and covered by it is a small, more primitive part of the brain, which developed very early in evolutionary history. This includes an extraordinarily important area known as the hypothalamus, which contains subcenters for control of circulation, body temperature, respiration and, among other functions, control of the anterior pituitary gland. The hypothalamus is not completely independent of the cerebrum; on the contrary, these two parts of the brain have many functional interconnections, and from time to time I will refer to the influence of the higher (cerebral) centers on the hypothalamus.

For many years scientists were convinced that the brain is involved in the control of the anterior pituitary gland, but how or by what mechanisms that control is exerted was not known. The fact that emotional factors in humans can have an effect on the function of the ovaries and the adrenal cortex, and the recognized influence of environmental factors, such as light, on female reproductive cycles in many lower animals indicated very strongly that the anterior pituitary must in some way be controlled by the central nervous system (the brain). Still another tantalizing observation was made long ago by those who study development of the embryo. The cells that ultimately form the anterior pituitary gland arise from an area in the embryo destined to become the mouth. However, these cells migrate toward a part of the embryonic brain that will form both the hypothalamus and the posterior pituitary gland. It is understandable that the very close anatomical association between the anterior pituitary and the hypothalamus would lead scientists to suspect that the hypothalamus must have something to do with the function of the anterior pituitary.

But at this point a very troublesome situation arose: It was soon found that the anterior pituitary has an extraordinarily scanty nerve supply; therefore, direct neural control seemed unlikely or even impossible. No such problem pertains to the posterior pituitary gland since it is formed from neural tissue in the embryo, is full of nerves, and is, in effect, an extension of the brain.

The first clues for resolving the dilemma of how the brain controls the anterior pituitary came about 1930, when it was discovered that between the hypothalamus and the anterior pituitary there is a distinctive network of small blood vessels. Later work showed that the principal flow of blood in these vessels is from the hypothalamus to the anterior pituitary gland, and this complex of blood vessels became known as the Hypothalamic-Hypophyseal Portal System, which we shall abbreviate to HHPS.

On the basis of these findings, it was evident that the hypothalamus might control the anterior pituitary gland by producing substances that are delivered directly to the anterior pituitary by way of the HHPS. The idea soon turned out to be correct; several hypothalamic compounds with profound influence over anterior pituitary activity have been isolated, and some of them have been synthesized in the laboratory. Thus, it is now clear that the hypothalamus acts as an endocrine gland that produces hormones having important effects on anterior pituitary function. In fact, it has been found that if the

HHPS is broken or cut so that hypothalamic hormones cannot reach the anterior pituitary, the anterior pituitary gland becomes almost functionless, with one exception—production of prolactin. With the hypothalamic connections all gone, the anterior pituitary gland stops production of all its hormones except PRL which is still poured out at a great rate.

In the next few paragrahs we will discuss how the brain, and more specifically the hypothalamus, acts like an endocrine gland. In a sense it works like any other endocrine gland in that it makes hormones that it delivers directly into the bloodstream. An interesting difference from other endocrine glands is that all the hypothalamic hormones are poured into the HHPS, which takes them directly to the anterior pituitary gland. In this way the anterior pituitary gets first crack at the hypothalamic hormones, and the rest of the body gets only what the anterior pituitary does not pick up. A similar situation pertains to the insulin-producing cells in the pancreas; in this case insulin is discharged into a part of the circulatory system that goes first to the liver, and the rest of the body gets whatever insulin the liver does not take first. However, the other endocrine glands discharge their hormones into general body circulation, and the hormones are then carried to all parts of the body in more or less equal concentration.

The role of the hypothalamus is not simply to control the menstrual cycle, and some hypothalamic hormones are not directly involved in the cycle. Current evidence suggests that the hypothalamus produces six different hormones, but the hypothalamus is still under very active exploration and it is possible that additional hypophysiotropic* hormones will be discovered. The easiest way to summarize our present knowledge of hypothalamic hormones is to present them in a list:

1. GRF—growth hormone–releasing factor.
2. GIF—growth hormone–inhibiting factor, also called somatostatin.
3. CRF—corticotropin–releasing factor (or adrenocorticotropin–releasing factor).
4. TRF—thyrotropin–releasing factor.
5. PIF—prolactin–inhibiting factor.
6. LRF—luteinizing hormone–releasing factor.

These names are fairly self-explanatory. All of those hypothalamic hormones that are "releasing factors" stimulate the anterior pituitary gland to secrete one of its hormones. For example, TRF induces release of TSH by the anterior pituitary; CRF causes ACTH secretion. The inhibiting factors have the reverse effect—they tend to turn off the release of the named anterior pituitary hormone.

You can imagine how exciting the discovery of hypothalamic hormones was to physicians and to scientists interested in human biology. A great deal of light was shed on hitherto obscure aspects of how our bodies function in both health and disease. Those studying human reproduction were particularly excited because the operation of the reproductive system began at long last to make some sense, although there is still a long way to go

*Acting on the hypophysis, or pituitary gland.

for a really detailed understanding of the relationship between the brain and the anterior pituitary gland. For now
we can certainly say that the anterior pituitary gland is
completely dependent on the brain for its normal operation.

Here we are particularly interested in the last three
hypothalamic hormones in the list—TRF, PIF, and
LRF. But why are we interested in TRF and PIF, since
neither thyrotropin (TSH) nor prolactin (PRL) seem to
be heavily or immediately involved in regulating the
menstrual cycle? It is true that they are not, but prolactin
is a very interesting hormone, and it does play an important part in reproduction. And TRF, which turns on release of TSH by the anterior pituitary, also excites prolactin release. This is a strange situation because, as we saw
above, prolactin is the one anterior pituitary hormone
that does not require a hypothalamic hormone or releasing factor for its output from the pituitary; it is secreted
without hypothalamic stimulation. The fact is we do not
know how or when TRF normally is used in controlling
prolactin output. Presently it is believed that the main
control is through the PIF mechanism, and when extra
prolactin is called forth, as during suckling, the anterior
pituitary gland increases its output because the
hypothalamus has reduced its release of PIF, which, you
will recall, is a prolactin-inhibiting factor. Less inhibition
on prolactin release permits the anterior pituitary to increase its output. Since PIF, an inhibiting hormone,
apparently represents the main control over prolactin release, you can understand why the anterior pituitary
gland secretes a great deal of prolactin when it is cut off
from hypothalamic control.

You might think of the anterior pituitary as having six faucets, each putting out a different hormone. All the faucets except the one for prolactin are normally closed, and the hypothalamic hormones are needed to open them and permit the anterior pituitary hormones to flow. The prolactin faucet is normally open, and prolactin will therefore pour out unless the hypothalamic PIF comes in to turn it off. At the time of suckling the hypothalamus reduces its output of PIF, and as a result the prolactin faucet opens wider, and prolactin flows at an increased rate.

At this point let us turn to the brain's action in controlling the production and release of milk shortly after parturition (birth of a child). Toward the end of pregnancy the blood levels of estradiol and progesterone, as well as of prolactin, are very high. The placenta,* the structure that is attached to the mother's uterus and to the infant's umbilical cord and is involved in transferring nutrients from mother to infant, produces most of the estradiol and progesterone at this time. It also augments the anterior pituitary's output of PRL by producing significant amounts of its own prolactin. The estradiol and progesterone from the placenta have been able to "prepare" the breast for milk production. Since PRL is also present, we might ask why milk is not produced even before the child is born. Nature has provided a deterrent in that the *very* high levels of progesterone, and possibly also estrogen, inhibit the actual production of milk. Since most of the estradiol and progesterone is of placental ori-

---

*The placenta is quite commonly called the afterbirth because it is discharged from the uterus shortly after delivery of the child.

gin, their concentrations in the blood fall very rapidly after the placenta has been expelled from the uterus of the mother. The deterrent to milk production is thus removed, and within a matter of hours the breasts begin to fill with milk.

At this point the hypothalamus and even the higher centers of the brain begin to play a fascinating role. Many a new mother is surprised and intrigued to find that, when the nurse brings her infant to be fed, there may be a spontaneous spurt of milk from the nipples brought about simply by the sight of her baby. This reaction obviously requires signals to the hypothalamus from other parts of the brain, and the hypothalamus in this case responds by transmitting a signal to the posterior pituitary gland. The posterior pituitary, in turn, releases a hormone called oxytocin, which reaches the breast by way of the bloodstream and causes certain cells around the milk-producing glands to contract and to squeeze milk into the ducts and out to the nipples. This is sometimes known as milk "let-down." Spontaneous milk let-down resulting from visual cues or from an emotional response to the infant illustrates the influence of the brain, but it is not a universal phenomenon. The same reaction practically always occurs when the child is permitted to suckle at the breast. In this case, stimulation of the nipple sends signals to the brain by way of sensory nerves, and this again results in the release of oxytocin from the posterior pituitary gland and then the release of milk. Two distinct actions are involved here: milk production and milk release. Milk production without milk release is not enough to make the milk available to the suckling infant.

Continued milk production requires that the breast continue to be stimulated by PRL; here again the controlling mechanism uses both neural and humoral means of communication. Nipple stimulation sends neural signals to the brain, which responds by permitting the anterior pituitary to release more PRL. The result is more milk production (but remember that the milk is not available to the baby without the second factor—oxytocin-induced let-down).

Now that the role of both prolactin and oxytocin has been defined, some additional items of interest should be mentioned. Many practices in our culture may have some influence on lactation, particularly the common use of alcohol and tobacco. Evidence has been presented to suggest that both may have deleterious effects on lactation. Nicotine has been found to inhibit prolactin release by the anterior pituitary, and alcohol is reported to inhibit oxytocin release by the posterior pituitary. Reduced prolactin could result in lower milk production, and less oxytocin could mean inadequate milk release. Therefore, a lactating woman who both drinks and smokes might expect to produce less milk and to release less of what she does produce.

Let us now consider a question that may have come to mind as you read this section on the hypothalamic hormones. Hypothalamic hormones control only five of the six hormones put out by the anterior pituitary, so in the list of hypothalamic hormones above you will find one that exerts some control over each of the anterior pituitary hormones except FSH. How about FSH? Is there no hypothalamic factor for separately regulating FSH output by the anterior pituitary? Apparently not—at least no one has yet found either a separate releasing factor or inhibiting hormone to control the FSH faucet in the anterior pituitary. However, it has been found that LRF turns on not only the LH faucet but the FSH faucet as well. This can be a tricky situation. There are times when the anterior pituitary puts out more FSH than LH, and other times when LH output is greater than FSH. With only one hypothalamic controlling factor, LRF, how does the anterior pituitary adjust the relative amounts of FSH and LH that may be required at any particular time? That is an interesting and provocative question, and we are just beginning to get at least partial answers. It seems likely that the ovaries tell the anterior pituitary how to adjust the ratio of FSH to LH in response to a given LRF stimulus, and the ovaries communicate by their output of estradiol and progesterone, both of which the anterior pituitary recognizes and is able to respond to. The hypothalamus also is sensitive to ovarian hormones and will adjust its production of LRF as it receives and recognizes estradiol certainly, and possibly also progesterone.*

We have now come full circle, having covered three important hormone-producing organs—the hypothalamus, the anterior pituitary, and the ovaries. The hypothalamus is regulated by ovarian hormones and higher brain centers; the anterior pituitary is dependent on the hypothalamus, and the ovaries are dependent on the anterior pituitary. The uterus and breasts are not endocrine glands —they do not produce hormones—but they do require hormonal support. All of these relationships are summar-

---

*Another ovarian hormone that inhibits FSH but not LH release has been found but not yet isolated. It is known as inhibin.

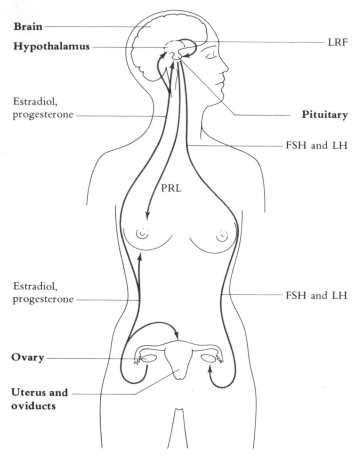

Brain

Hypothalamus

LRF

Estradiol,
progesterone

Pituitary

FSH and LH

PRL

Estradiol,
progesterone

FSH and LH

Ovary

Uterus and
oviducts

FIGURE 2.4

An outline of the body of a woman showing the
positions of the various parts of the menstrual system.
Boldface labels indicate structures; roman labels
indicate hormone actions. The arrows leading from one
structure to another indicate hormones. The tips of the
arrows point to the structure or structures influenced
by each hormone; the other ends of the arrows are at
the sources of the hormones. Compare with Figure 3.1.

ized in Figure 2.4 and will be considered in more detail in the next chapter.

In a sense, the discussion in this chapter has moved backward, starting with the uterus and working back (or up) to the brain. I did it this way because I thought it might make more sense if we started at the end with the uterus and then got some idea as to how each step of the process depends on a preceding one. It would have been difficult and a little awkward to discuss the effect of the anterior pituitary on the ovaries if the ovaries had not been discussed first, or the influence of the hypothalamus on the anterior pituitary if a discussion of the pituitary had not preceded the one on the hypothalmus. Now it is time to put all these bits and pieces together and to move forward. An overall coherent picture of the menstrual cycle, how it is regulated, and how that regulation sometimes breaks down to cause menstrual disturbances are the subjects of the ensuing chapters.

3

PUTTING
IT ALL
TOGETHER

some general principles . . . .

This will undoubtedly be the most technical chapter in the book, but that should not be cause for concern. Read this chapter more slowly than the others and spend some time with the diagrams as you read. Don't expect to retain a detailed picture of the operation of the menstrual cycle when you complete the chapter. So long as you come away with a general idea of how the menstrual cycle works, of the relationships among the various parts of the cycling system, then don't worry about the details. You can always check back to refresh your memory on specific items if that should become necessary.

In Chapter 2 we discussed the separate or independent functions of the uterus, the ovaries, the anterior pituitary gland, and that part of the brain known as the hypothalamus. Now we will attempt to assemble these four different structures to form a working system we will call the *menstrual system*. Obviously, none of these four components of the system can by themselves create a menstrual cycle, but when they are put together in a healthy body where they can interact normally, one of the results is a menstrual cycle. The menstrual system, like most systems either created by human beings or designed by nature, has characteristics that are difficult or impossible to predict from a knowledge only of their separate parts. It is the relationship—the interaction—between and among those parts that yields a system with a unique set of functions not possessed by any of the components alone.

The generally accepted principle is that in a working system the whole is equal to more than the sum of all its parts. Poincaré, a renowned French mathematician, perhaps said it best when he wrote in 1902: "Science is built with facts, as a house is with stone. But a collection of facts is no more science than a heap of stones is a house." Consider, for example, a familiar system—a television set. It is made up of many different kinds of parts: transistors, condensers, resistors, wire, and so on. None of those parts alone will behave like a television set; nor will a sack full of all the parts of a television set. But when those parts are assembled so that they can interact appropriately, we get fully operating television.

In the menstrual system, as in the television set, each individual part performs its normal function only when it is able to interact with the other parts of the system. When all those parts are properly put together and allowed to function, the result is the monthly release of a tiny ovum, ready to be fertilized, and the monthly renewal of the tissue that will cradle it if it should be fertilized. Not only does this system far surpass a television set in the complexity of its operating parts, but it also possesses a kind of beauty and significance not assignable to any system made by human beings.

As we have seen, the hypothalamus is the master controller for the menstrual system. The hypothalamus does receive signals from other parts of the brain, probably those parts dealing with vision, with the emotions, and with thought; and its function is also influenced, as we will see, by its exposure to estradiol and progesterone from the ovaries. However, the hypothalamus appears to be able to operate without such input and thus is not dependent, in the way the ovaries are completely dependent on the anterior pituitary and the anterior pituitary is dependent on the hypothalamus.

The great dependence of the entire reproductive system on the hypothalamus is exemplified by its role in determining when puberty will occur. As we saw in Chapter 2 in the discussion of breast development, this sign of approaching sexual maturity requires the hormones produced by the ovaries. We might wonder why the ovaries wait to produce their hormones until a girl is, typically, around 11 or 12 years old. In earlier times it was believed that the ovaries had to mature along with the rest of the body. However, it was later recognized that the ovaries are ready to operate much earlier than they actually do and that they are merely awaiting the necessary signals— the gonadotropic hormones FSH and LH—from the anterior pituitary. So we must now ask why the anterior pituitary holds off until the age of puberty. Some experimental evidence indicates that the anterior pituitary is also ready to operate long before the age of puberty, but it is waiting for the appropriate signal from the hypothalamus. It seems, then, that the ultimate responsibility resides in the hypothalamus, and the start-up of the system at puberty awaits the maturing of this part of the brain. Exactly what the maturation process is and how it is dependent on other parts of the brain is not now known. The exciting thought here is that the brain is responsible not only for controlling the menstrual cycle but also for determining when the menstrual life of a girl is to begin. Think of it for a moment: It is the brain, or at least some part of it, that determines when a girl's breasts will develop, when her pubic hair will begin to appear, when she will begin to menstruate.

A rather puzzling situation arose several years ago when it was found that the sex-related changes associated with puberty seem to begin before measurable increases in gonadotropic hormones can be detected. Remember that the changes associated with puberty require ovarian hormones in girls and a hormone from the testicles in boys, and gonadotropic hormone from the anterior pituitary in both cases. Quite recently it was discovered that the required increase in gonadotropic hormone was taking place, but it was being secreted by the pituitary during sleep. Up to that time, all the measurements had been made during the day in boys and girls who were awake, and the sleep-related increase in anterior pituitary activity therefore was not detected. Undoubtedly, the sleep-associated gonadotropic hormone output is dependent on the hypothalamus, so it appears that the hypothalamus awakens for LRF output when the young boy or girl goes to sleep, and sex development associated with puberty thus usually begins at night.

You have already seen that the four parts of the menstrual system operate in a well-defined sequence. In a sense there is a kind of "domino effect," with the hypothalamus stimulating the anterior pituitary, which then activates the ovaries, which in turn produce hormones having an effect on the uterus. This chain of events is a fundamental part of the menstrual system, but of itself it will not induce repeated cycles. There is one additional and extremely important ingredient, and that is an influence of the ovarian hormones on the first two parts of the system, the hypothalamus and the anterior pituitary. For obvious reasons, this effect is known as feedback control, and this part of the system is frequently called a feedback loop. What are the consequences of ovarian feedback control? We know that the hypothalamus and anterior pituitary determine the activity of the ovaries, but by way of the feedback loop the ovaries also influence the hypothalamus and anterior pituitary. So we have a kind of circular system in which the ovaries have considerable control over their own function.

Figure 3.1 provides a graphic illustration of the functional relationships among the various parts of the menstrual system and of the feedback loop. Examine this figure carefully in order to become familiar with the sys-

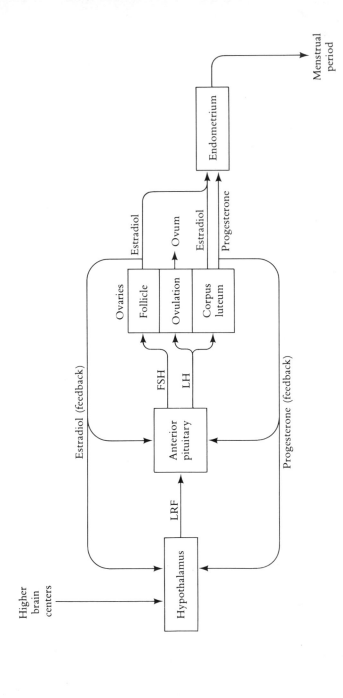

tem and the way it operates. This diagram borrows a bit from the engineers and from those biomedical scientists who are involved in systems analysis. We are, in a modest way, trying to analyze a system, so this kind of model is appropriate. This diagram displays in a different form the same information presented in Figure 2.4.

Each of the boxes in the diagram represents one of the components of the menstrual system discussed in Chapter 2. Note that each box has one or more arrows going in and one or more going out. The arrows going in are known as inputs, while those arrows pointing away from the boxes are outputs. Input and output are conventional terms but they mean about the same as cause and

FIGURE 31

A diagram of the menstrual system. This diagram shows the four parts of the menstrual system discussed in Chapter 2: the brain (represented by the hypothalamus), the anterior pituitary, the ovaries, and the uterus (represented by the endometrium). The relationships among these parts of the system are also illustrated. Note that LRF, the output of the hypothalamus, is the input to the anterior pituitary. The resulting outputs of the anterior pituitary are FSH and LH, which are inputs to the ovaries. FSH causes follicular development; LH induces ovulation and corpus luteum formation. The ovarian hormones estradiol and progesterone are outputs of the ovaries, and they influence the endometrium and also provide feedback control of the anterior pituitary and hypothalamus. Compare this figure with Figure 2.4.

effect or stimulus and response. The amount of output in each case is dependent on the amount of the associated input and on how much or how little the response may be amplified by the box itself.

The diagram can also be seen as a kind of flow chart for the menstrual system. An important feature of all systems is the fact that the output of one component of the system frequently serves as the input to another. Note that in the case of the menstrual system LRF, the output of the hypothalamus, is the input to the anterior pituitary, and FSH and LH, both outputs of the anterior pituitary, are inputs to the ovaries. This was discussed in Chapter 2, but the diagram illustrates it.

LRF can be thought of as a signal transmitted by the hypothalamus and received by the anterior pituitary. Obviously, with no signal there would be no response in the receiver. As in the case of radio transmission, the stronger the LRF signal, the stronger is the pituitary response. But the strength of that response is also determined by something else: the feedback control by estradiol and progesterone. These two ovarian hormones determine the magnitude of the response of the anterior pituitary to any given LRF input, and they also have an influence on the LRF output by the hypothalamus. It is important to understand feedback control because it is commonly used to stabilize and regulate the systems designed by engineers, and it is also a method frequently employed by nature to do the same for biological systems. In the case of the menstrual system, the feedback effect of the ovaries is required for the menstrual system to go through repetitive cycles. Without that feedback loop, or when the feedback signals are abnormal, menstrual cycles may stop.

Let me illustrate the importance of feedback by describing a familiar example of a system in which feedback is significantly involved. A thermostatically controlled home heating system has two essential components, a thermostat and a furnace. The thermostat is simply a temperature-sensitive switch that signals the furnace when heat is required. Thus the input to the furnace is the signal from the thermostat; the output of the furnace is heat. That heat is also the feedback signal to the thermostat. When the heat warms the thermostat to a given point, the thermostat turns off its signal to the furnace, and the furnace shuts down. The furnace will not turn on again until the heat loss from the house lowers the temperature of the thermostat, which then signals the furnace to turn on. So, in response to the thermostat, the furnace will turn on and off at intervals; in other words, it will cycle. Suppose that the heat—the feedback signal to the thermostat—were to be discharged outside the house so that the thermostat would not receive it. Now there is no feedback to the thermostat, so the furnace will run continuously; in other words, the cycling will stop.

A similar, although much more complicated, arrangement pertains to the menstrual system. Without the ovarian feedback signals, the menstrual system also stops cycling. We will have occasion to discuss feedback regulation in somewhat greater detail as we consider the operation of the menstrual system and when the subject of menopause is introduced. Contraceptive pills, as we will see in Chapter 6, function by taking advantage of feedback principles.

In general, estradiol from the ovary tends to suppress the release of both LRF from the hypothalamus and FSH from the anterior pituitary. Consequently, when estradiol is absent, as in a woman whose ovaries have been removed, the hypothalamus produces excessive amounts of LRF, and the anterior pituitary secretes vast amounts of gonadotropic hormone—more FSH than LH. If estradiol is injected into such an individual, her hypothalamic and anterior pituitary activities are diminished. So the principal feedback effect of estradiol, as we have seen, seems to be to inhibit the release of gonadotropic hormones; because it *reduces* the activity of the hypothalamus and anterior pituitary, the effect is known as *negative* feedback.

How do the separate actions of the parts of the menstrual system work together to create a menstrual cycle? The rest of this chapter should help to provide an answer to that question. However, many aspects of the control of the menstrual cycle are not yet understood even by those medical scientists who are working at the frontiers of our knowledge, so some features of the cycle may not be completely clear. For those parts of the operation of the menstrual system about which we may have little or no direct information, I will frequently provide a kind of educated guess* based on whatever indirect evidence may be available.

In reviewing the menstrual cycle, we must agree on how we will count the days. Let us assume that we have a cycle 28 days long. By convention, the first day of a menstrual period is counted as the first day of the cycle. We may want to object to this because, in a sense, the

---

*Such guesses may, of course, turn out to be wrong; but I agree with Francis Bacon, a seventeenth-century English philosopher, who is said to have once remarked that "truth will sooner come out of error than out of confusion."

menstrual period is the end of a chain of events, not the beginning; but to avoid confusion we will accept the conventional way of counting. Besides, if we are dealing with a truly cyclic system, it should not make much difference where we break into it to start counting.

The usually accepted divisions, or phases, of the menstrual cycle are illustrated in Figure 3.2. Again, we are assuming a 28-day cycle, day 1 of which coincides with the first day of a menstrual period. The menstrual period usually lasts about five days and is followed by a preovulatory period known as the follicular, or estrogenic, phase. The rationale for these terms will become evident as we discuss the control of the cycle. Ovulation in a 28-day cycle usually occurs on about day 14, and this leads into the so-called luteal, or progestational, phase of the cycle, which lasts another 14 days, after which the next menstrual period begins.

As has been mentioned several times, the total length of the cycle may vary considerably, and 28 days merely represents an average. A rather curious feature of the cycle, however, is the fact that the length of the (postovulatory) luteal phase tends to be relatively constant—about 14 days—regardless of the length of the entire cycle. Consider this for a moment and examine Figure 3.3, which illustrates the duration of the various parts of the cycle for three different cycle lengths. Note that menstrual periods are specified in each case as lasting about five days, but, more significant, the luteal phase is constant at about 14 days. This means that ovulation always occurs about 14 days before the beginning of the next menstrual period, no matter how long the entire cycle may be. Suppose we consider a 25-day cycle. When

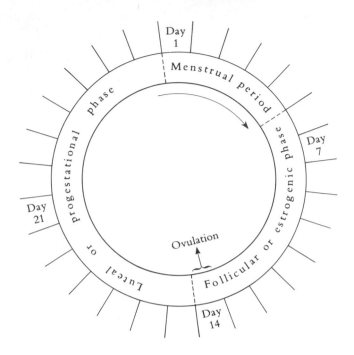

FIGURE 3.2

The phases of the menstrual cycle. By convention, day 1 of a menstrual cycle is defined as the first day of a menstrual period. The menstrual period, which lasts for about five days, is followed by the follicular, or estrogenic, phase of the cycle, which in turn is followed by the progestational, or luteal, phase of the cycle. When the cycle is 28 days long, as indicated in this diagram, ovulation usually takes place on day 14. See Figure 3.3 for more on the timing of ovulation.

would ovulation occur and the luteal phase begin? Fourteen days before the next menstrual period would be about day 11, and, as is shown in Figure 3.3, that is approximately the day ovulation would occur. For a cycle

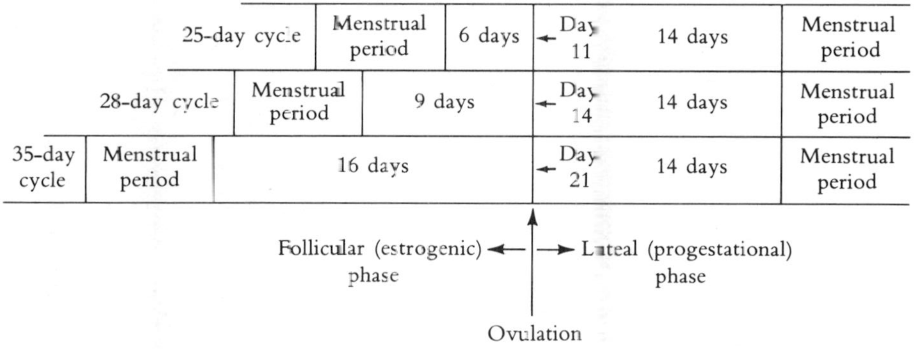

25-day cycle	Menstrual period	6 days	Day 11	14 days	Menstrual period
28-day cycle	Menstrual period	9 days	Day 14	14 days	Menstrual period
35-day cycle	Menstrual period	16 days	Day 21	14 days	Menstrual period

Follicular (estrogenic) phase ⟵——— ———⟶ Luteal (progestational) phase

Ovulation

FIGURE 3.3

The influence of the length of the menstrual cycle on the various phases of the cycle and on the time of ovulation. The luteal phase of the menstrual cycle remains constant in length, while the follicular phase of the cycle varies when the length of the entire cycle changes. Ovulation marks the end of the follicular phase of the cycle.

longer than 28 days, the luteal phase is still about 14 days long; as Figure 3.3 shows, the follicular phase is the phase that changes as the whole cycle varies in its length. *If* we knew exactly how long a particular menstrual cycle was going to be, we would be able to predict with fair, but not complete, reliability the day on which ovulation was due to happen. However, the length of an ongoing cycle is difficult to predict, except perhaps for women whose cycles are very regular or constant in length. Even then the time of ovulation, although predictable on the average within a day or two, cannot be predetermined with certainty. We will have occasion to discuss this matter again as it relates to the issue of birth control.

the first half of the cycle: the estrogenic phase . . . .

As we take up the events of the menstrual cycle, refer to the diagrams presented in Figures 3.1 through 3.4. Let us begin with a menstrual period, which, on the average, lasts from day 1 through about day 5 of the cycle. At this time the ovaries are not very active, and estradiol and progesterone levels are consequently quite low—not as low as in a woman without ovaries, but low enough so that the negative feedback effect of estradiol on the hypothalamus is diminished and, as a result, hypothalamic LRF in substantial amounts begins to be released. The anterior pituitary responds by secreting a relatively large amount of its gonadotropic hormone, consisting mainly of FSH. This process actually begins late in the previous cycle and is in full swing by about day 3, during the time of a menstrual period. LH is not extensively released because it appears that when the estradiol level is low, the anterior pituitary favors the secretion of FSH over LH. One of the two ovaries responds to the high concentration of FSH in the blood by developing one of its follicles, which contains an ovum and which, under the continuing influence of FSH, begins to secrete estradiol (see Figure 3.4).

Now we encounter a most intriguing problem. The two ovaries seem to take turns in producing mature follicles. With each menstrual cycle they alternate, so that in any one cycle one ovary is busy preparing a mature follicle and its ovum and producing hormones, while the other ovary rests. In the next cycle they switch roles, and the resting ovary gets to work, while the ovary that had produced a follicle, released an ovum, and formed a corpus luteum in the previous cycle now rests. So it goes, right ovary, left ovary, right ovary, left ovary, and so on, in alternate menstrual cycles. The problem is how an ovary "knows" when to respond to the hormones from the anterior pituitary and when those stimulating hormones can be ignored. No one knows for sure what the mechanisms are or how that control is exerted.

The mystery deepens in a woman who has had to have one of her ovaries removed. The remaining ovary seems to recognize that it is the only ovary left, and instead of resting every other cycle, it takes over the job of the departed ovary and goes through the entire process of follicle formation, ovulation, and corpus luteum formation every month. How does that hardworking ovary "know" that it is alone and must respond to pituitary stimulation with each cycle? Again, no one knows.

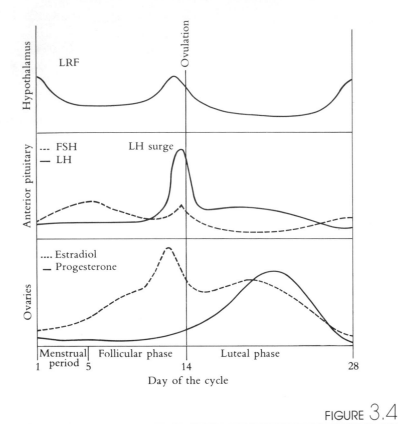

FIGURE 3.4

Changes in hormone levels during the menstrual cycle.
Note particularly the peaks of estradiol and LRF, which
provide the LH surge.

Let us consider some other ovarian puzzles. Suppose
we focus our attention on the one ovary that is re-
sponding to FSH from the pituitary. The ovary has
thousands upon thousands of ova. How does it "decide"
which one to select for the maturation process? How does
it manage to keep the other ova from also maturing and
being released? Why in human females is only one ovum

usually released—occasionally two, and rarely three or more—while in the rat, for example, both ovaries with each cycle will release several ova? Here again, we must plead ignorance; we don't know what the control system is or how it works.

In recent years there have been more frequent reports of women giving birth to quadruplets and quintuplets. In almost all these cases the women involved have received hormone treatment that has overstimulated the ovaries so that multiple follicles, probably from both ovaries, have matured and discharged their ova. This, of course, represents an interference with the normal operation of the menstrual system; on the other hand, the treatment was undoubtedly given because the system was not operating quite normally anyway.

Let's return to the story of the first half of the menstrual cycle. We left that story when one of the two ovaries was developing a follicle and producing increasing amounts of estradiol, all under the influence of FSH from the pituitary, which was being pushed by LRF from the hypothalamus. All that estradiol coming from the ovary has a number of important effects. First of all, it causes a rapid growth of the endometrium. Don't forget that during the previous menstrual period the endometrium had been stripped down to a very thin layer, usually called the basal layer. But now, under estradiol's influence, the endometrial cells rapidly divide to form a thick endometrium again. Estradiol also has an effect on the uterine muscles (the myometrium), which during this part of the cycle begin to exhibit contractions at a relatively high frequency; but they are rather mild (low amplitude) contractions, and women are not aware of them.

All these processes continue to about day 13 of the menstrual cycle. By that time the endometrium, under the stimulating influence of continuously increasing estradiol levels, has become quite thick. The kind of endometrial growth seen at this time has a distinctive appearance, and a skilled pathologist looking at a small piece of such endometrium under the microscope can say without hesitation that this bit of endometrium was taken during the first half of a menstrual cycle. Appropriately, it is called an *estrogenic endometrium*. Also by day 13 the "working" ovary has produced a well-developed follicle, perhaps about half an inch or more in diameter and standing out at or near the surface of the ovary, sometimes looking like a blister (see Figure 2.2). It should now be easy to see why the part of a menstrual cycle that follows the menstrual period is usually called the follicular phase or, sometimes, the estrogenic phase.

While all this is going on, estradiol begins to exert a profound feedback effect on the hypothalamus and the anterior pituitary. Evidence currently available suggests that LRF output by the hypothalamus is then reduced; this may simply be negative feedback in action. The estradiol effect on the anterior pituitary is just a bit more complicated since it appears to have several simultaneous effects:

1. Estradiol increases the sensitivity of the anterior pituitary to LRF.
2. Estradiol also causes the anterior pituitary to emphasize LH production over FSH production.
3. However, estradiol inhibits the *release* of that LH by the anterior pituitary.

Now let's look at the results of these three actions of estradiol. The increased sensitivity (item 1) means that the anterior pituitary continues to make gonadotropic hormones at a relatively high rate in spite of the decreased LRF stimulus. Most, though not all, of the gonadotropic hormone so produced is in the form of LH (item 2), but most of the LH made at this time is stored in the anterior pituitary rather than released (item 3). All these processes continue to about day 13 of the menstrual cycle. By this time the anterior pituitary is loaded with LH.

Now the stage is set for some dramatic happenings involving the hypothalamus, the LH-laden anterior pituitary, and the mature follicle poised at the surface of the ovary. However, the events to follow may be considered part of the next scene in this drama, so let's go on.

ovulation . . . .

Scientists are particularly tantalized by this part of the menstrual cycle because the controlling mechanisms for events I am about to describe are not clearly understood. Some possibilities have been presented, but we still don't know exactly how the system works at this special time of the cycle.

The main issue has to do with how the menstrual system recognizes that the time has come for ovulation. Again, the hypothalamus is in charge, and its release of extra LRF is believed to be required. So the question is, What input to the hypothalamus triggers the release of additional LRF at this time? Although we don't have a clear-cut answer to that question yet, we can explore some of the possibilities. Estradiol may be involved; Figure 3.4 shows that prior to ovulation the estradiol level reaches a peak concentration in the blood and then falls very rapidly. It appears that the hypothalamus is able to respond to the very high estradiol concentration by releasing additional LRF.

Here is an unusual situation. We said that earlier in the follicular phase estradiol inhibits LRF release by direct negative feedback. Now we are suggesting that estradiol at very high levels *stimulates* LRF release. These two situations appear to be in conflict, but the blood concentration of estradiol may make the difference. When the estradiol level is very low, LRF output is high, and increments of estradiol will suppress LRF release. This is relatively commonplace negative feedback. On the other hand, when estradiol levels are already fairly high but going higher, the effect on LRF secretion reverses, and now the increases in estradiol concentration in the blood seem to stimulate LRF release. This is sometimes called a positive feedback action of estradiol, in contrast to its usual negative feedback. This situation is illustrated in Figure 3.5. As a general rule, positive feedback leads to instability and may have strange effects on a system. At this particular time of the cycle, however, that positive feedback may be necessary for achieving the events soon to follow. At any rate, estradiol is certainly involved, and it has been found that administration of estrogen in the preovulatory period can, in fact, induce ovulation.

Another possibility includes a proposed role for progesterone. Recent evidence has shown that the blood

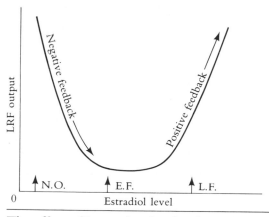

FIGURE 3.5

The effect of increasing levels of estradiol on LRF output by the hypothalamus. Note that LRF output is high when the estradiol levels are either very low or very high, thus yielding a U-shaped curve. The small arrow labeled N.O. indicates the conditions when no ovarian function is present; the other two small arrows, E.F. and L.F., show the conditions during the early and later follicular phases, respectively. The decreasing output of LRF on the left side of the curve results from the negative feedback of increasing levels of estradiol; the rising output of LRF on the right is caused by the positive feedback of still higher levels of estradiol.

level of progesterone may begin to rise a day or so prior to ovulation (see Figure 3.4). This rise seems to occur only in a menstrual cycle in which ovulation takes place,* and it has been suggested that the preovulatory increase in progesterone may somehow also be involved in signaling the hypothalamus that the time is ripe for ovulation.

*Cycles in which ovulation does not occur will be discussed later in this chapter.

Unfortunately, we are not able to describe more definitively the input to the hypothalamus at this critical time. Changes in estradiol, in progesterone, or in both may be required, but regardless of the nature of that input, the hypothalamus responds by releasing a large amount of LRF. The menstrual system reacts in a special way at this particular time. The anterior pituitary, now full of LH and especially sensitive to LRF (because of estradiol influence), suddenly gets that large dose of LRF from the hypothalamus and responds in a most dramatic manner. The anterior pituitary's reaction can be likened to a gigantic paroxysmal sneeze, with a sudden and convulsive release of an enormous amount of its stored LH. Small wonder that the result is widely known as the LH surge. The LH surge is shown in Figure 3.4, which also indicates that FSH is released in considerable quantity at the same time. The FSH release appears to have no particular significance and may simply reflect the fact that the anterior pituitary is unable to secrete large amounts of LH without releasing FSH too. The internal pituitary mechanisms for release of the two gonadotropins may be similar.

Let's turn our attention back to the ovary with the mature follicle. It stands ready for the flood of LH released at the time of the surge, and about 12 hours later the follicle reponds by breaking open and discharging its ovum and follicular fluid (see Figure 2.3). Just how LH causes the follicle to rupture is not known precisely. At one time it was thought that growing internal pressure of follicular fluid simply ruptures the follicle. But that is not the full explanation. Apparently as a result of some local chemical action, a tiny area of the follicular wall begins to weaken, and the pressure of the follicular fluid pushes this

spot outward so that it takes on the appearance of a mini-
ature nipple. That little spot begins to ooze follicular fluid
and ultimately opens wide, rapidly releasing more fluid,
which also carries the ovum out. The ruptured wall of the
follicle may also bleed a little at this time, resulting in the
mittelschmerz described in Chapter 2. All this happens
on about day 14 of the cycle, which is a day of special
significance—the day of ovulation. In Figure 3.1 you will
see that LH is specified as leading directly to ovulation

In Chapter 1 we mentioned animals that do not ovulate spontaneously. The rabbit and the cat, for example,
usually ovulate only following the stimulus of copulation
(sexual intercourse). In this case the sex act stimulates sensory nerves in the area of the vagina; these neural signals
travel up the spinal cord to the brain and rapidly arrive at
the hypothalamus, which responds by releasing additional
LRF. That LRF very soon reaches the pituitary and induces an LH surge which, several hours later, causes the
mature ovarian follicles to burst and release their ova. As
indicated in Chapter 1, this system of ovulation does not
pertain to humans; women ovulate spontaneously and
more or less on schedule, regardless of their sexual
activity.

The released ovum is discharged from the follicle
into the abdominal cavity, but it doesn't wander around
because very close to the ovary is the funnel-shaped opening (called the infundibulum) of the Fallopian tube (the
oviduct). The cells lining the inside of the tube have little
hair-like structures, known as cilia, which beat in an or-
ganized pattern like thousands of tiny oars, and the fluid
in and around the tube begins to flow in a direction that
helps to bring the ovum into the oviduct and keeps it

from getting lost in the abdominal cavity. Thus the likelihood that an ovum will lose its way and fail to find the opening of the oviduct is very low. Of course, the ovum does not have far to go because, as I mentioned before, the infundibulum is very close to the ovary. However, there are reported cases of women who have been able to get pregnant even after having one ovary and the Fallopian tube on the opposite side removed. This means that, if necessary, an ovum can find its way from the ovary that released it to the Fallopian tube on the opposite side, a fairly long trip.

the second half of the cycle: the progestational phase . . . .

Just before ovulation occurs, the estradiol output by the ovary decreases markedly. This causes a substantial, although temporary, drop in estradiol level, which is sometimes of sufficient magnitude and duration to represent a significant loss of hormonal support to the endometrium. Don't forget that during the first half of the menstrual cycle estradiol has been responsible for building up the endometrium, and if the estradiol level should fall, that endometrium may begin to break down again. As I say, that sometimes happens about the time of ovulation, and a little bit of bleeding, or "spotting" may occasionally show up.

Under the influence of continued release of LH, although at a much lower rate than during the surge, the ruptured ovarian follicle is converted into a corpus luteum, which was described in Chapter 2. A corpus luteum is formed only if ovulation has occurred because the corpus luteum is built on the ruins, so to speak, of a ruptured follicle. However, once formed, the corpus luteum begins secreting estradiol so that the blood level of this hormone goes up again; in addition, the corpus luteum begins the release of large amounts of progesterone (see Figure 3.4).

Progesterone has a profound effect on the uterus, both on the myometrium and on the endometrium. As was mentioned in Chapter 2, there are spontaneous contractions of the muscles of the uterus not only at the time of the birth of a baby but also during the course of an ordinary menstrual cycle. During the first half of the cycle, the follicular, or estrogenic, phase, there are many mild contractions (high frequency, low amplitude). With the addition of progesterone in the second half of the cycle, the contractions continue, but they change in character; they occur less frequently but are much stronger (low frequency, high amplitude).

The endometrium also responds to the progesterone. It becomes very thick and rather spongy, and develops many mucus-secreting glands. This is the kind of endometrium required if an embryo is to attach and to develop. Because this part of the menstrual cycle is under the predominant influence of the corpus luteum and progesterone, it is known as the luteal, or progestational, phase of the cycle. The endometrium is also classified as progestational, and microscopic examination can distinguish between an estrogenic and progestational endometrium.

Since a corpus luteum is formed only after ovulation, the presence of a progestational endometrium means that ovulation has taken place. Gynecologists sometimes want to know what the endometrium looks like and will remove a tiny piece for examination. If it is a progestational endometrium, they can be quite certain that ovulation has occurred. This can be important information in cases in which pregnancy is desired but for some reason does not occur. Another way to tell whether ovulation has occurred is by taking a small sample of blood and analyzing it for progesterone level. We now have remarkably sensitive methods for analyzing all the hormones we have discussed.

During the second half of the menstrual cycle—that is, during the luteal phase—the high level of progesterone, in addition to its profound effect on the endometrium, makes its presence known in a number of other ways. As mentioned earlier, both estradiol and progesterone have an effect on the breasts. The presence of both these hormones during the second half of the cycle causes some enlargement and engorgement, and as a result the breasts may become somewhat sensitive and possibly even painful. The breast changes at this time are similar to those found during early pregnancy, but naturally they are of shorter duration and lesser degree.

However, the changes are those that prepare the breasts for lactation, and you might wonder why lactation does not occur during the progestational phase of each menstrual cycle. I suppose that we are fortunate that it doesn't because that might be a bit inconvenient and certainly useless. Fortunately, the prolactin levels are not high enough to induce significant production of milk.

Since nipple stimulation is a good way to induce the anterior pituitary to secrete additional prolactin, you might ask what nipple stimulation would accomplish toward the end of the luteal phase of a menstrual cycle? Can a woman who has not recently delivered a child produce any milk? In fact, it is possible, and it has happened that under such circumstances a drop or two of milk is expressed. Although this is an interesting phenomenon, in most cases the experiment would fail—not enough estradiol and progesterone and for too short a time. The prolactin production probably is also inadequate in both amount and duration.

Still another change may sometimes be observed in synchrony with ovulation and the second half of the menstrual cycle. Recall that the lower end of the uterus, the cervix, protrudes slightly into the vagina, and through it runs a channel (the cervical canal) that connects the main body of the uterus with the vagina. Spermatozoa deposited in the vagina at the time of sexual intercourse must make their way through the cervical canal into the uterus and from the uterus into the Fallopian tube, where fertilization takes place. The cervical canal always contains mucus, which forms a sort of plug, and that mucus plug goes through cyclic changes correlated with the menstrual cycle. During most of the cycle the mucus is rather thick and viscous, and spermatozoa find it somewhat difficult (but not necessarily impossible) to penetrate.

However, around the time of ovulation the secretion of cervical mucus increases markedly, and the mucus itself becomes more fluid. There may even be some leakage from the vagina. Spermatozoa can get through this mucus

with relative ease. Following ovulation the mucus becomes viscous again, is secreted in smaller amounts, and
is not pentrated by spermatozoa very readily. These
changes in the character of cervical mucus are sometimes
used to determine the time of ovulation. This is discussed
in Chapter 6.

Let us return to the second half (luteal phase) of the
menstrual cycle. Through the major part of this segment
of the cycle, the estradiol and progesterone secretion by
the corpus luteum is quite high, and toward the middle or
end of the luteal phase the LH level begins to fall. The
reasons for this change are not clear, although a negative
feedback effect of the high levels of progesterone has been
suggested (see Figure 3.1).

Regardless of the mechanism involved, when LH
falls, the corpus luteum, which is dependent on LH, begins to degenerate, and its production of estradiol and
progesterone rapidly falls. This happens on about day 27
and 28 of the cycle, and all the effects we have ascribed to
progesterone are now reversed. For example, the breasts
decrease in size again and are no longer painful; but the
most obvious and dramatic effect is that the endometrium
begins to break down.

How does the endometrium "know" that the progesterone level is falling? Its receptor sites for this hormone are not being adequately supplied. But what happens
exactly when the progesterone receptors in this fully developed endometrium are inadequately provided with
hormone? By some mechanism certain blood vessels that
supply the endometrium with blood begin to constrict,
reducing the amount of blood flowing to this part of the
uterus. When that amount of blood is not sufficient to

keep the endometrium supplied with nutrients and oxygen, the endometrium begins to deteriorate. Pieces of endometrium fall away from the inner wall of the uterus, breaking some of the blood vessels, which, even though they are constricted, still ooze a little blood. The bits of endometrium and the blood, mixed with some mucus and water, finally begin to flow through the cervix into the vagina—and day 1 of the next cycle has begun.

One of the early studies of menstrual mechanisms involved a rather unusual approach. In the 1930s an American investigator named J. E. Markee decided that he wanted to be able to observe the endometrium during the course of a menstral cycle, so he removed a tiny piece of endometrium from the uterus of a monkey and transplanted that bit of endometrium into one of the monkey's eyes, just under the transparent cornea and off to one side, where it did not interfere with vision. The transplant developed blood connections with the eye and was easily studied by bringing a microscope up close to the cornea, through which the endometrium was clearly visible. Markee reasoned that the endometrium in the eye should go through the same changes as the endometrium in the uterus and that in the eye he would be able to see those changes. He was quite successful and observed that late in the luteal phase of the cycle, the endometrial transplant would turn pale periodically, indicating that blood flow was being reduced from time to time. Markee called this the "blush-blanche" phenomenon, and he noticed that as the cycle progressed, the intervals of blanching or blood flow reduction increased both in frequency and duration until, toward the end of the luteal phase, long periods of blanche occurred. Finally, because of reduced blood supply, bits of the little piece of transplanted endometrium began to break down, and a kind of mini-menstrual period took place right in the eye. This may seem a bizarre kind of experiment, but Markee was a pioneer in discovering some of the mechanisms involved in menstruation.

cycles without ovulation . . . .

Before leaving this analysis of the menstrual cycle, we should mention cycles in which ovulation does not occur. Such cycles are known as anovulatory cycles and are probably rather rare during most of the menstrual life of a woman, but they do occur frequently at the beginning and the end. In fact, the first few cycles in the girl at puberty and the last few in the woman approaching menopause may all be anovulatory. Superficially, an anovulatory cycle cannot be distinquished from an ovulatory cycle. They may be of the same duration, and the menstrual flow may be similar for the two. On the other hand, remember that in the absence of ovulation a corpus luteum is not formed, and consequently progesterone is not produced. In the absence of progesterone, all the usual effects described for the luteal or progestational phase of a menstrual cycle are absent. The breasts do not become painful, and the endometrium remains in an estrogenic form rather than becoming progestational.

An anovulatory cycle may occur either because of an inadequate LH surge, normally required to induce ovulation, or because the ovaries are unable to respond properly to LH. In the case of pubescent girls, the former is probably the cause; in menopausal women it is very likely the latter.

In either case the estrogenic phase of the cycle may occur normally, but in the absence of ovulation this phase of the cycle is extended into that period when a luteal phase would normally have occurred. Eventually the negative feedback effect of the maintained high level of estradiol reduces the output of FSH by the pituitary; the ovarian follicle deteriorates, and the estradiol level then begins to fall. This means that the endometrium loses its hormonal support, begins to break down, and a menstrual period then takes place.

As I stated earlier, many, many features of the control of the menstrual cycle are not yet understood. We are dealing with an exquisitely regulated and incredibly complex arrangement, and there are many gaps in our knowledge. These gaps are rapidly being filled, and in the next few years I anticipate clarification of many aspects of the menstrual cycle about which we are still in the dark.

However, these first chapters have been intended to give some understanding of how the various parts of the menstrual system work together and of the inputs responsible for the changes occurring in the ovaries and uterus during the course of a menstrual cycle. They have, I hope, illustrated the beautiful intricacy of the system, the highly coordinated way in which the components function as a unit, and the role of the hypothalamus, which, aided both by the higher centers of the brain and by the ovarian hormones, manages to keep the menstrual system doing its job month after month.

4

WHY THE MENSTRUAL CYCLE SOMETIMES STOPS

As mentioned in Chapter 1, most women have menstrual cycles for about 35 years, starting perhaps at about 12 or 13 years of age and stopping at about 48 or so. The ages for puberty, the beginning of menstrual life, and menopause, its irreversible termination, will vary from one individual to another and also from one part of the world to another. For example, girls in the very warm or tropical parts of the world may start menstruating at a somewhat earlier age than those living in the temperate or cold climates, and they usually stop menstruating at an earlier age, too.

For the most part, during the 35 years or so of their menstrual life, women will cycle more or less regularly. The cycles may or may not be of equal duration; for a given woman the cycle length may vary from time to time, and certainly it may vary rather widely among individual women. It is rather unusual, although not unheard of, for a menstrual cycle to be much shorter than 20 days or much longer than 38 or 40 days. This is the case during most of the reproductive history of normal women, but their cycles may be erratic and of much greater length both at the beginning and the end of their menstrual life.

Although most women will cycle fairly regularly most of the time, occasionally certain situations do arise that may interfere with the menstrual cycle or even stop it. The most common and obvious of these situations is pregnancy. Everyone knows that pregnant women do not menstruate; thus, the onset of a menstrual period is regarded as a sign that pregnancy has not occurred.

pregnancy and lactation . . . .

Although the absence of menstruation, or amenorrhea, during pregnancy is common knowledge, the reason for the cessation of cycles is not so widely known. Actually, the mechanism is rather simple in concept. Toward the end of a menstrual cycle in the absence of pregnancy, the corpus luteum in the ovary, which has been supported by LH from the anterior pituitary, begins to deteriorate. This ultimately results in a decrease in output of estradiol and progesterone, on which the endometrium depends; in the absence of adequate estradiol and progesterone, the endometrium breaks down and a menstrual period ensues. If pregnancy intervenes, an altered set of circumstances is set up. Let us review what happens.

First of all, the ovum, discharged from the ovarian follicle on the day of ovulation (about day 14 of the cycle), finds its way into the oviduct (or Fallopian tube), and, if sexual intercourse has occurred at the appropriate time, it is met by large numbers of spermatozoa. Fertilization of the ovum by a single sperm cell (Figure 4.1) usually happens within about 24 hours or so after ovulation. In fact, it is generally believed that the ovum does not remain fertilizable much longer than about 24 hours after it has left its ovarian follicle.

Many men and women think that fertilization takes place in the uterus, but this is incorrect. Ordinarily, the ovum is fertilized while it is still in the oviduct, and it begins to divide to form an embryo while it is still there. It takes about 5 days or so for that dividing ovum to make the trip down the oviduct into the uterus, and when it gets there, the endometrium—right in the middle of its progestational phase—is ready to receive it. The dividing ovum, by this time a small ball made up of many cells and known as the blastula, embeds itself deep in the endometrium—all of this about seven days after ovulation, or on about day 21 of the cycle.

During the next several days certain embryonic membranes, called chorionic membranes, are formed, and within a few days these begin to function like endocrine

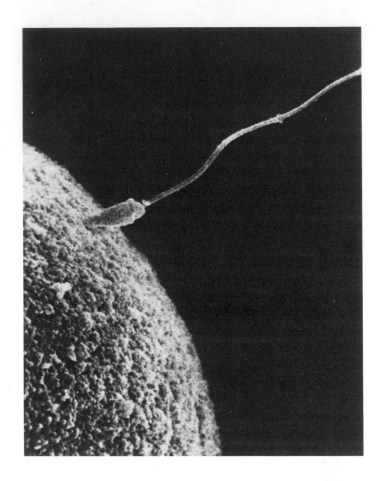

FIGURE 4.1

The moment of fertilization. This dramatic picture was taken with an instrument known as a scanning electron microscope. It is magnified many thousands of times over actual size and shows a single sperm cell in contact with and just beginning to penetrate an ovum. These are not human cells; they were taken from a marine animal, the sea urchin, which is commonly used for studying the process of fertilization. (Photograph by Dr. D. W. Fawcett and Dr. E. Anderson of Harvard Medical School. Reprinted with permission.)

glands, producing and secreting hormones. One of these hormones, known as human chorionic gonadotropin (HCG), is released in increasing amounts beginning on about day 27 of the cycle. Remember that this is the time when the endometrium is beginning to suffer because the corpus luteum is not getting enough LH to keep its estradiol and progesterone output up. But HCG, which acts almost exactly like LH, comes along just in the nick of time to rescue that corpus luteum and keep it going. Because of the continued release of HCG, the corpus luteum persists; the resulting maintained release of estradiol and progesterone preserves the endometrium and the embryo it contains—and, of course, the scheduled menstrual period does not occur. Some of the HCG in the pregnant woman passes through her kidneys and then appears in her urine in substantial quantities. Most of the tests used to determine the presence of an early pregnancy are based on the finding of HCG in the urine, where it can be detected immediately after the first missed menstrual period.

The high HCG output continues for close to three months and then diminishes rapidly, but by that time the placenta is also producing large quantities of its own estradiol and progesterone, and the endometrium is no longer dependent on the ovary and its corpus luteum for support. Production of estradiol and progesterone by the placenta continues at increasing levels throughout the remainder of pregnancy, and thus the pregnant woman continues to be amenorrheic (without menstrual periods) until after the child and placenta are delivered. With the loss of the placenta, the source of estradiol and progesterone is gone, and the endometrium, no longer supported, then breaks down.

Another cause of delayed or irregular menstrual periods is lactation. When the mother nurses her baby for some period of time, menstrual periods may be delayed. The apparent reason is the high level of prolactin from the anterior pituitary, although we do not know for sure how prolactin operates to interfere with normal cycling. In some animals prolactin acts something like LH and thus tends to maintain a corpus luteum and its output of estradiol and progesterone for as long as the suckling stimulus keeps the prolactin levels up. If this were the case in humans, the continued high levels of estradiol and progesterone would maintain the endometrium and prevent a menstrual period. Futhermore, by way of the negative feedback loop, the estradiol and progesterone would inhibit the release of gonadotropic hormones by the anterior pituitary, and thus an ovarian cycle would also be prevented. However, it is not clear that this mechanism operates in nursing mothers, and some other effect of prolactin, that enigmatic hormone, may be involved.

Regardless of the mechanism, lactating women do not cycle regularly. Because of this it is widely, but erroneously, believed that nursing mothers never ovulate and therefore are not fertile. Actually, lactating women can and do ovulate occasionally and thus can become pregnant.

menopause . . . .

Menopause is the other obvious time when menstrual cycles stop. Usually the cessation of cycling is not abrupt; the cycles may become a bit irregular for several months before they stop altogether. In addition, as was mentioned before, the last few cycles may be anovulatory.

The cause of the cessation of cycling in menopause does not reside in either the hypothalamus or in the anterior pituitary gland, both of which appear to be functional indefinitely. Instead, the ovaries become unresponsive to stimulation by either FSH or LH from the anterior pituitary, and as a result they regress. No one knows precisely why the ovaries ultimately lose their sensitivity to gonadotropic stimulation, although it has been found that the blood vessels of the ovaries do tend to deteriorate about the time of menopause, which of course, would interfere with proper transport of gonadotropic hormones to the ovaries. We are then led to ask why the ovarian blood vessels decline. I know of no certain answer to that question, but no matter what the cause, with the ovaries not functioning estradiol and progesterone levels in the blood will fall. With the feedback effect of estradiol removed and the hypothalamus and anterior pituitary still normally operational, LRF and FSH are released in large quantities, as was described in the previous chapter, but the ovaries are unresponsive.

emotional causes . . . .

Of all the many factors that can influence the menstrual cycle, emotional and social factors may be the most intriguing. The menstrual cycle can be modified or even stopped by various emotional upsets, by change in occupation, by moving to a different part of the world, and by many other psychological and environmental factors.

This observation is not based on isolated and uncon-
firmed stories of the kind that most of us have heard
at one time or another. Critical studies have been made
demonstrating quite conclusively that the higher centers
of the brain can indeed have significant impact on the
menstrual cycle, and on lactation as well. This fact is
perhaps best illustrated by various kinds of emotional disturbances that result in a cessation of menstrual cycles.

Amenorrhea sometimes occurs in women who are
convinced that they are pregnant, although they really are
not. This condition is commonly known as false pregnancy, sometimes as phantom pregnancy; the technical
term is *pseudocyesis*.* Here we have a remarkable situation, in which a woman is so certain she is pregnant that
her menstrual cycle stops and menstrual periods are
consequently postponed. You can readily see how
pseudocyesis is an example of the profound effect that the
higher centers of the brain have on reproductive function,
undoubtedly through an immediate action on the function of the hypothalamus.

Just how those parts of the brain that control mood
and emotion influence the hypothalamus is not clearly
understood, but this matter is under very active investigation right now. Certain chemical compounds produced in
the brain may be required by the hypothalamus for its
normal function. These compounds, as a group, are
known as catecholamines, and they include adrenalin and
related substances. Certain tranquilizers that tend to block
the action of catecholamines are also known to cause

---

*Cyesis is derived from a Greek word meaning pregnancy;
pseudocyesis, then, is false pregnancy.

amenorrhea and prevent ovulation. Thus it is reasoned that catecholamines from the higher brain centers may represent an important link between those centers and the hypothalamus. An inappropriate catecholamine input to the hypothalamus could then result in inadequate LRF stimulation of the anterior pituitary, with the ultimate consequence of no ovarian cycle, poor ovarian hormone production, and no menstrual cycle. At any rate, this chain of events may at least partly explain how emotional disturbances may interfere with the normal menstrual rhythm.

Now let us consider some examples of the impact of the emotions on the menstrual cycle by describing some cases of pseudocyesis. A relatively common background to the amenorrhea of pseudocyesis has been the fear of pregnancy in unmarried girls. Note that I use the past tense; I suspect that this basis for pseudocyesis has regressed with the greater availability of effective birth control methods and, if necessary, of abortion. More important, perhaps, is the markedly decreased social disapproval of premarital and extramarital sexual experience. In any case, the example that follows took place several years ago.

A 19-year-old girl came to see a gynecologist because her menstrual period was about two weeks overdue. They sat and talked quietly for a little while, and the young woman finally confessed that she had had sexual intercourse for the first and only time about three weeks before, and she was terribly afraid that she might be pregnant. Because of her family and religious background, she was especially upset and guilt-ridden. The gynecologist examined her thoroughly and found no signs of pregnancy. He attempted to reassure her and told her that she would very likely menstruate within a few days. About ten days later the girl returned, still not having menstruated and now convinced that she was pregnant. After further examinations and tests again no signs of pregnancy were found, and again the girl was sent home with the promise that she would shortly menstruate. Less than two weeks later, now having missed a second period, she appeared in the office again, tearful and frantic. The gynecologist performed another pelvic examination and then told the desperate young woman a lie: "You are just beginning to menstruate." She went home, and her menstrual period began that night.

Sometimes the kind of amenorrhea I have just described occurs in women who desperately want to have a child and can't seem to get pregnant even after trying for a long time. Many such cases over the years have appeared in the medical literature; they are now being studied with modern methods of hormone analysis. A recently published example describes a young woman who came to the obstetrician about eight months after she had stopped menstruating. She had no doubt that she was pregnant and was very pleased. She had experienced "morning sickness," her breasts were enlarged as in pregnancy, she claimed to feel movement of the baby, and her abdomen was distended as though she were about eight months pregnant. Measurements of her hormone levels were made, and it was found that the LH and prolactin levels were higher than normal. Her progesterone level was also elevated. However, a physical examination showed that her uterus was of normal nonpregnant size and that no embryo was present; intestinal gas was found to be responsible for the abdominal distention. When the physician finally convinced the woman that she was not really pregnant, the intestinal gas was passed, and her distended abdomen returned to normal, nonpregnant dimensions in about 30 minutes. About one week later she menstruated, and all the signs of pregnancy disappeared. Here is another case in which the higher centers of the brain have influenced the reproductive system, undoubtedly by way of the hypothalamus.

Fear of or desire for pregnancy are not the only psychogenic bases for amenorrhea. A phenomenon called college amenorrhea or boarding-house amenorrhea sometimes occurs in girls when they leave home for the first time and go off to college or a job. The emotional upset, apparently caused by leaving a secure and protected home environment for the first time, seems to be the precipitating factor. The unexpected loss of menstrual periods can, of course, be worrisome, but sympathy and reassurance frequently will start the cycles going again.

Still another kind of psychogenic amenorrhea is known medically as anorexia nervosa. *Anorexia* means a loss of appetite for food, and the *nervosa* indicates that it is of nervous origin. How this is related to the menstrual cycle will become evident in a moment. This disease, which can be very serious, has been known for a long, long time. It occurs most commonly in adolescent girls, many of whom report having had upsetting sexual experiences, and all of whom, either through loss of appetite or through deliberate dieting, may lose weight to the point of emaciation. There appears to be little doubt that anorexia nervosa is a psychiatric disease and that the emotional upset, whatever its basis or background may be, causes both the loss of appetite and also an abnormal function of the hypothalamus, which results in cessation of the menstrual cycle.

I have already introduced catecholamines as the link between the higher centers of the brain and the hypothalamus. It appears that in the anorexia nervosa the underlying emotional disturbance causes changes in the production of catecholamines such that the hypothalamus does not function properly, and without a properly operating hypothalamus, the menstrual system of course breaks down. Treatment of anorexia nervosa might involve several approaches, but psychotherapy may be the most important and successful means of treating this disease.

We could fill an entire book with descriptions of various kinds of emotionally induced changes in the menstrual cycle. The few examples I have given should make clear that the cerebrum, that part of the brain that provides awareness and mood, can play a significant role in the control of the menstrual cycle. As explained before, it seems evident that these higher regions of the brain exert their influence by modifying the actions of the hypothalamus. So even though the hypothalamus is a kind of master gland dominating the anterior pituitary, and through it the ovaries also, it does not act with complete independence or without influence from outside itself. Earlier, we discussed at some length the feedback effects of estradiol and progesterone on the hypothalamus. Now we see that there are also pathways for control from the higher centers of the brain.

defects in the system...............

A wide variety of defects may cause the menstrual cycle to stop or to behave in an erratic way. However, I do not intend to discuss or even list all of the problems that can befall the menstrual system. Considering the incredible complexity of the system and its control, the frequency of problems is remarkably low. For most women the system operates very well indeed. However, as might be expected, if any part of the system begins to misbehave, the entire system may be affected and may even come to a halt. In the previous section we discussed emotional causes for the cycle's stopping. Assuming that the basic cause in such cases resides in the higher centers of the brain, we can, in a sense, say that a "defect" in the system has occurred there. More frequently, however, the defect is found within the hypothalamus, the anterior pituitary, or the ovaries. A substantial battery of tests is now available to determine just exactly where the problem is. Obviously, if the ovaries do not respond properly to the gonadotropic hormones FSH and LH, or if the ovarian feedback signals to the hypothalamus and anterior pituitary are inappropriate, the cycle may become erratic or even stop.

On the other hand, even if the ovaries are completely normal, defects in the anterior pituitary or hypothalamus may stop the cycle. Terms such as *hypothalamic amenorrhea,** for example, have invaded the medical literature.

At any rate, even though it may not happen very often, it is reasonable to expect that in the menstrual sys-

*The cessation of menstrual cycles due to a hypothalamic defect.

tem, as in any complex system, a defect in any component, no matter how small or obscure, can cause the entire system to break down. But let me emphasize once again that in the vast majority of females, once the cycle begins, it continues more or less regularly for about 35 years or so, possibly stopping only for pregnancy and lactation.

5

ARE WOMEN REALLY DOMINATED BY THEIR MENSTRUAL CYCLES?

All of us are influenced by our physical and cultural environments. Although the word *environment* usually suggests an external set of conditions, we are also subjected to what is frequently called an internal environment. The continuously changing chemistry of our bodies represents that internal environment, and the hormones we have been discussing are significant parts of that environment.

I think that no one would argue with the statement that mood and behavior can be profoundly influenced by the internal environment, just as social and cultural factors as components of our external environment have long been recognized to exert such influence. That statement is equally applicable to both males and females of any age. However, the menstrual system with its intricate controls brings to women a cycling internal environment that has some impact on the brain as well as the uterus. Men, too, have cycles that may involve the hypothalamus and the anterior pituitary, although the cycles so far detected in the male are not as obvious as those in the female and probably not as regular. I will refer to male cycles again later in this chapter.

Let us return to the situation in women. Assuming that the cycling internal environment does influence the brain, particularly those parts of the brain that determine disposition, conduct, and so on, we may wonder whether this is another example of feedback. In Chapters 2 and 3 we discussed the influence of the higher centers of the brain on the cycle; now we see that the cycle may effect the higher centers of the brain. Although this is an interesting possibility, a well-defined feedback loop of this sort has not been identified. Nevertheless, regardless of whether it is or is not part of a feedback mechanism, the menstrual cycle does seem to have cycling influence on those parts of the brain that affect disposition, mood, behavior, and personality.

A few words of caution with regard to studies on this subject are appropriate here. First of all, the methods available for measuring characteristics of mood or behavior may in some cases be open to question, and the results produced by these methods consequently may be difficult to interpret. Secondly, we must be careful not to extrapolate by drawing conclusions about the cyclic behavior of women beyond what objective observation and experience would logically permit. I have already referred to one study with questionable methods, a study dealing with whether or not sex drive, or libido, in women varies cyclically with the menstrual cycle. One method of measurement is simply to ask the women who serve as subjects for the study to record the frequency of sexual intercourse during various parts of the cycle. Although there is little doubt that libido in women will influence the frequency of intercourse, libido is not the only operative factor and thus may not determine how often coitus occurs. For example, the desires or sexual needs of a woman's mate may certainly play an important role. Cultural and social milieu, economic status, and level of education are all variables that influence sexual practices.

When faced with a complex situation of this sort, the investigator must make a choice. Either these variables can be controlled by restricting the study to a well-defined and circumscribed segment of the population and conclusions drawn only for that group of women, or the variables must be randomized by using a large and unbiased sample of subjects. You can see why it is that inconsistent observations are sometimes made, and why the results obtained by different investigators may not be in agreement. I don't want to leave you with the idea that all such studies are poorly made or not worthwhile. On the contrary, such work is very important, and many such investigations—difficult as they are to design—have been well conceived and well executed. But interpretations of results and the methods that produced them must be critically evaluated before they are either accepted or rejected.

Enough of this editorializing. Let us now consider two possible mechanisms that may be involved in the influence of the menstrual cycle on mood or behavior. For a long time it was commonly believed that during the progestational phase of the menstrual cycle, considerable excess water is retained in the body and stored in various organs, with some of it accumulating in and causing a swelling of the brain. This was and still is believed by many to be partly responsible for the behavioral changes that are thought to occur in the second half of the cycle. However, evidence now available makes it no longer possible to conclude with confidence that such changes in water balance are responsible for cyclic changes in mood and behavior.

Another possible mechanism involves a direct effect of certain hormones, particularly the ovarian hormones, on the brain. Certain parts of the brain possess specific receptor sites for estradiol and progesterone, as well as for other hormones, and a hormone-receptor complex usually does not exist unless it has some function to perform. At any rate, ovarian hormones are picked up and presumably "used" by the brain, and they may thereby have a significant influence on personality characteristics.

premenstrual problems . . . .

Regardless of the cause, there is no doubt that large numbers of women experience some discomfort during the three or four days just preceding a menstrual period.

Surveys have concluded that about 60 percent of women are affected, but for most of them the symptoms are quite mild, including perhaps a headache, a feeling of fullness, sometimes an increased irritability and insomnia, occasionally slight depression with perhaps a tendency for tears, and so on. This collection of symptoms is commonly known as premenstrual tension or the premenstrual syndrome.

Syndrome is a Greek word meaning simultaneous occurrence; in English it is commonly defined as a group of concurrent symptoms. That definition is frequently extended to stipulate that the symptoms involved are characteristic of some abnormality or disease. For this reason we could object to calling the premenstrual condition a syndrome. Why should we think of the premenstrual changes as an abnormality? Those changes are a common feature of the menstrual cycle, and certainly the cycle is not a disease.

In any normal function of the body, exaggerated or depressed reactions or adjustments may sometimes indicate an abnormality. For example, an elevated heart rate and heavy breathing are normal responses to exercise. However, if the heart rate and breathing should go too high after a given exercise, this may suggest disease, possibly of the heart or lungs. A decision of how high is too high requires judgment and a knowledge of the range of responses in a normal population. Similar statements can and should be made about premenstrual symptoms. There are some women, fortunately not many, in whom serious symptoms can include profound depression and other forms of emotional instability. Educated judgment may determine when the symptoms are so severe as to represent an abnormality. However, for the vast majority of women, premenstrual discomfort must be regarded as a common feature of a normal function.

You may perhaps object to the idea of classifying any physical or mental discomfort as normal. However, we humans encounter many, many situations that typically lead to discomfort, and we don't think of that discomfort as a symptom of disease. For example, noise in the external environment can lead to insomnia, and this is considered to be an unfortunate situation but not a sign of disease. Premenstrual insomnia might be said to result from a "noisy" internal environment and should not be arbitrarily labeled abnormal.

In spite of the common occurrence of premenstrual symptoms, little is known about their cause, and there are differences of opinion about treatment. As mentioned earlier, it was widely thought at one time that water retention (edema) in the brain might be the cause, but forcing the loss of excessive body water through the use of drugs (diuretics) usually does not alleviate all of the symptoms. Claims have been made that giving orally effective progesterone-like compounds (progestins) will reduce premenstrual symptoms, and it has been suggested, therefore, that the symptoms result from an inadequate release of progesterone during the luteal phase of the cycle. Despite many suggestions, no standard and effective treatment for premenstrual tension has yet been devised.

A psychological factor also appears to play a significant role in the development of premenstrual symptoms. In a recently published study of over 40 women, an interesting technique was described for convincing them that they were in stages of their menstrual cycle that differed from the actual cycle phase. Those women who were led to believe that they were in the premenstrual period reported having greater symptoms than those women who thought that they were in the middle of the cycle. These results do not mean that the premenstrual symptoms are without a physical cause, but they strongly suggest a significant psychological component as well.

painful periods — dysmenorrhea

Another difficulty that besets large numbers of women is dysmenorrhea—painful menstrual periods. In such cases the first day or two of the period are usually the most difficult in that they include a particular kind of abdominal or menstrual cramp, which can be so severe as to be quite debilitating. Strangely enough, dysmenorrhea seems to occur only in ovulatory cycles and can be alleviated by a hormonal treatment that creates anovulatory cycles. Such treatment is usually administered only in very severe cases of dysmenorrhea. Many women find that an aspirin tablet or two takes the edge off the pain; others believe that certain exercises or postures may help. Some women report that menstrual cramps subside after intercourse, particularly if the sexual experience has led to orgasm.

The underlying cause of menstrual pain is not known with certainty, although several theories have been presented. The following causes have been proposed:

1. An inappropriate ratio of estradiol to progesterone in the late luteal phase of the cycle.
2. Failure of the endometrium to break down completely.
3. Blockage of the canal through the cervix to the vagina.

Many other presumed causes too numerous to list completely have been considered. However, one factor that may play a role in production of menstrual pain has to do with the kind of uterine contractions occurring in the premenstrual and early menstrual periods. Strong and sustained contractions of the muscles of the uterus have been found just before and during menstruation; in the case of dysmenorrhea, these contractions may be even more intense. As I mentioned earlier, the uterus contracts more or less throughout the menstrual cycle, so why should strong contractions at the time of the menstrual period be painful? The answer seems to lie in the pattern of contraction. When any muscle is made to work with an inadequate blood and oxygen supply (ischemia), great pain results. In fact, such pain, known as ischemic pain, may be one of the most severe we can experience—most of us have had a muscle spasm or cramp in the leg that attests to this fact. Ischemia is also the origin of the pain in a heart attack, when one of the arteries to the heart is plugged by a clot and the heart is made to work with less than the necessary amount of blood. Menstrual cramps are sometimes considered to be an example of ischemic pain. In this case the muscle contractions may be so strong as to reduce blood supply simply by squeezing down on the blood vessels. As long as the spastic contraction persists, the uterine muscle is working with an inadequate supply of blood and oxygen, and the severe pain results.

The strong contractions of the uterus during the first day or so of a menstrual period may be caused by the increased production in the uterus of some chemical compounds known as prostaglandins. Prostaglandins are commonly found in various parts of the body, and one of their effects is to cause the contraction of the kind of muscle that makes up most of the uterus. Certain drugs that inhibit the production of prostaglandins have been developed and are being tested for their ability to diminish or eliminate the pain sometimes associated with a menstrual period. In fact, this may be partly the way in which aspirin diminishes such pain; aspirin has been found to be one of the drugs that interfere with the production of prostaglandins.

As pointed out earlier, a majority of women experience some premenstrual symptoms and menstrual discomfort, but the range of severity is great. Not only is there an endocrine basis for this broad range of symptoms, but there also appears to be a significant psychological component. Well-adjusted, happy, and active women on the average seem to have less premenstrual difficulty than do those who are dissatisfied, bored, or depressed in general. Similar statements can be made for the severity of dysmenorrhea.

It is believed by some that the severity of premenstrual and menstrual symptoms may, in part, be determined by how a young girl is prepared for her menstrual function. If she is told to expect 35 to 40 years of monthly misery and that menstruation is indeed a curse on women, if her mother is irritable and goes to bed for a few days each month, she may be more likely to experience menstrual difficulties. On the other hand, if the prepubertal girl is taught that menstrual cycles are simply a fact of life for women, if she knows the biological significance of the cycle, if she has been shown how to care for herself during a period and perhaps how her mother does it, then the probability is greater that her menstrual life will not be a burden. Let me state also at this point that I firmly believe that all the menstrual information given to girls should also be made available to boys. If both sexes understand the menstrual function of women, then perhaps the social, vocational, or cultural discrimination that ignorance has permitted to flourish will be eliminated, or at least minimized, and the menstrual period will not remain the "curse" many women now regard it to be.

menopause

There is another circumstance that may also have emotional consequences. This is menopause, the time when menstrual cycles come to an end. Much of what I have said about the premenstrual and menstrual days pertains also to the transition into menopause. Many women experience both physical and emotional symptoms, but again the severity ranges widely.

As indicated earlier, the cessation of ovulation and the decline in estradiol secretion by the ovary is a gradual process that may take several years. Menopause itself is defined as the cessation of menstrual periods; the years just before and after the menopause are known as the pre-menopausal and postmenopausal periods, respectively. Although the age for menopause may vary considerably, for most women it occurs sometime between 47 and 52. At this time the ovaries become unresponsive to the anterior pituitary hormones, and, as a result, hormone production by the ovaries dwindles down to a very low level.

Among the most common symptoms in postmenopausal women are the so-called hot flushes and night sweats. These can be quite disagreeable but are subject to treatment. Earlier I mentioned that the hypothalamus has many important functions other than its hormonal control of the anterior pituitary. Among these functions is some regulation of the circulation of blood, and one way the hypothalamus exerts this effect is through its role in controlling the diameter of many small blood vessels. This is known as a vasomotor effect. During the menstrual life of the woman, her hypothalamus apparently becomes accustomed to the presence of much estradiol. This can be considered to cause a sort of estradiol addiction, and when the ovaries stop producing estradiol, the hypothalmus exhibits withdrawal effects that are manifested as a vasomotor instability. In other words, because of estradiol withdrawal and under the resulting inappropriate orders of the hypothalamus, the blood vessels, particularly of the face and neck, will from time to time dilate and produce a flushing of the skin and an uncomfortable sensation of heat. Other parts of the brain may also be involved in this reaction.

Menopause can involve other symptoms, some of which may be emotional, including depression, nervousness, and inability to concentrate. To some extent such symptoms are probably produced by the estradiol deficiency of menopause, but social or cultural factors may play a role also. Even though their menstrual periods may have brought discomfort, many women are quite upset at the time of menopause when their periods stop. If women are fearful of advancing age, or if they have been led to believe that they will suffer a loss of femininity, that their sex lives will deteriorate, that their mental equilibrium is in danger, and that purpose and reason for continued existence is diminished or gone, then understandably menopause will bring depression.

For those problems that are directly related to estradiol deficiency, the solution is relatively simple. One needs only to replace the lost estradiol with a similar substance. Estradiol itself is not very effective when taken by mouth because, after being absorbed by the intestines, it passes through the liver, where it is inactivated. However, a number of orally effective estrogenic compounds are available, the most commonly used being a rather poorly defined group of compounds extracted from the urine of the mare.

Little doubt exists that estrogen replacement will reverse the physical effects of menopause, but considerable controversy persists concerning the possibility of potential danger. The presumed danger has to do with the relationship between estrogens and cancer of the endometrium, the cervix, and the breast, with the endometrium and cervix regarded as being at greater risk. This is an issue that has yet to be resolved. Many oncologists (cancer specialists) claim that we have no conclusive evidence that estrogens cause cancer in the women who take these hormones; others, however, feel that the use of estrogens may predispose a woman to endometrial cancer or at least accelerate the development of such a cancer if it were to occur. These are matters of great concern and importance and are under active exploration; definitive answers may be forthcoming within the next few years. For the time being, most gynecologists seem to feel that the benefits that can be derived from small amounts of needed estrogen far outweigh the presumed risk.

What about the social factors that predispose women to some of the unpleasant symptoms of menopause? These, of course, are more difficult to deal with, but knowledge of the biology of menopause can be helpful. As was indicated above, menopause is simply that time of a woman's life when her ovaries stop working. The ovaries no longer discharge ova, nor do they produce much, if any, estradiol and progesterone; and menstrual periods come to a halt. With the loss of estrogen particularly, a number of physical changes may be observed; for example, skin texture may change, the breasts may "deflate" somewhat. But of utmost importance is an understanding that every woman and man should achieve: Menopause does not mark the end of a useful and productive life; it does not mean that a woman is suddenly less capable of discharging her responsibilities, either professionally or at home; it is not the end of her ability to both take and give pleasure in a sexual relationship; and it does not mean she will suddenly lose her femininity and physical attractiveness.

Many women, particularly those with active interests, have very little difficulty in making the transition from the menstrual years to the postmenopausal years. From time to time they may experience some of the physical symptoms, but these can be controlled through appropriate use of small amounts of hormones if they are severe enough to be upsetting. Similarly, women who have been sexually active during their reproductive years and who have not been brainwashed into believing that menopause brings a loss of sexuality, or that their mates will no longer find them sexually attractive, for the most part continue to be sexually active for an indefinite period. In fact, some women report that sex improves postmenopausally as contraceptive measures are no longer required. With the ovaries no longer working, the fear of an unwanted pregnancy is, of course, eliminated.

On the other hand, we must not neglect or discount the fact that with a complete loss of estradiol, the lining (mucosa) of the vagina, which is estrogen-dependent, may eventually undergo regressive changes that can render sexual intercourse uncomfortable or even somewhat painful. This does not happen immediately or to all postmenopausal women, because the amount of change in the vagina is dependent on the degree of estrogen loss, and the estrogen level after menopause varies considerably in different women. Should this become a problem, it can be resolved by taking small amounts of prescribed estrogens.

Students of human sexuality claim, however, that even in the absence of adequate estrogen, continued or regular intercourse will tend to preserve at least some of the vaginal mucosa and prevent or minimize the discomfort that might otherwise occur if sex were very infrequent. Naturally, none of this applies to women in their cycling years since the ovaries then produce more than enough estradiol to maintain the lining of the vagina.

Let us consider here one last social or cultural factor. Most of the Western cultures place a high value on youth and, conversely, downgrade middle and old age. We are bombarded by propaganda in newspapers and magazines and on television on the great virtues of youth; we are beset by advertisements on methods and products for maintaining at least a youthful appearance. Many young people regard the idea of sex for those in their fifties and sixties with some distaste—possibly as a kind of obscenity. In some segments of our society, its older members seem to represent a kind of embarrassment, and they are pushed aside, hidden in "rest homes," or—perhaps even worse—merely ignored. Under such circumstances it is small wonder that menopause, if it is equated with old age, should be disturbing. I am told that in some oriental cultures where the aged, particularly the older women, are venerated and have special status within both the family and the community, menopause is usually not a problem, and emotional responses to this stage of life are virtually absent.

We must nevertheless be realistic about advancing age and what it means. Few if any of us like to dwell on the fact that we are mortal and that life therefore has both a beginning and an end. However, menopause should not be equated with declining mental and physical vigor; it should not connote a dried up, sexless old woman. Many men and some women look at menopause as the beginning of the end. That notion is unreasonable. Menopause is a stage of life, and although it usually occurs when life may be more than half over, the remaining years, like the earlier years, have their advantages as well as their disadvantages. We cannot deny the physical changes that may occur postmenopausally. However, rich and rewarding years are still available to the postmenopausal woman.

the question . . . .

Now let us return to the title of this chapter and ask, "Are women really dominated by their menstrual cycles?" The word *dominated* has been used deliberately, even though the idea of such domination may be offensive to most women. However, this idea is so prevalent (mostly among men, I suspect) that the question must be asked and answered. The concept of monthly domination and its implication of periodic instability has, in part at least, been responsible for women's being denied access to certain kinds of opportunities that have traditionally been available to men only. Thus it is an issue of great importance.

Some of what you are about to read is a reflection of personal opinion. This is not scientifically inappropriate, because scientists do much more than simply observe; they also interpret their observations. Furthermore, my opinions will not always be (and have not always been) stated dispassionately. Here again, no apologies are made because scientists, who are after all human beings, can exhibit excitement and enthusiasm without being guilty of creating bad science.

Now, to the question. Certainly it is a fact that women have well-defined cycles that may involve cyclic changes in mood or disposition. However, I know of no evidence that the average healthy woman is incapacitated emotionally or not to be trusted with social or professional responsibilities. Those with extraordinarily painful periods can get relief, and they should seek appropriate medical intervention.

Of course, there may be a small minority of women whose emotional stability may be borderline to begin with and whose ability to handle responsibility may deteriorate premenstrually. For example, it has been reported that over 40 percent of those women entering hospitals for psychiatric illness are admitted during the critical interval just before and during the menstrual days. More women are arrested premenstrually than at other times of the cycle; however, some investigators believe that women do not commit more crimes at this time but rather that their ability to avoid detection may be reduced. Many such observations exist, some well and systematically made and others largely anecdotal. Moreover, such problems exist equally among men. Although most men, like most women, are stable and trustworthy, some may run along the borderline of emotional stability and occasionally go over the edge. Furthermore, there is accumulating evidence that men probably also have cycles, not obvious monthly cycles analogous to menstrual cycles, but cycles nonetheless. In a recent study that involved measurements of emotional state and behavior patterns in a fairly large number of married couples, the men were found to exhibit as much monthly variability as the cycling women. I do not believe that this provocative finding reflects adversely on the human condition, and consequently I reject the interpretation that men are no better off than women. On the contrary, since variation can be an enormously important asset to all of us, I am pleased to conclude that in this case men have the same advantage as women.

I cannot understand why we human beings should want to deny our variability, be it cyclic or random. Variation is a characterstic common to all living things—all animal life, all plant life. Variation is essential for progress; without it life would be dull indeed. Happily, we are not machines with monolithic functions and fixed responses. Sometimes we are depressed, and sometimes we are elated. We vary, we change, we even cycle; because of all this—and not in spite of it—we laugh, we cry, we advance, and we are or can be creative, both biologically and intellectually.

I find somewhat irritating the apparent need to defend the female half of the human population. It seems gratuitous, and even embarrassing, to have to state that healthy, well-adjusted men and women are equally qualified for intellectually and emotionally demanding positions. Discrimination, in this case as in most cases, appears to be based on misinformation, irrational interpretation of the facts, and prejudice. Do men and women have differences? Of course they do. Besides the obvious anatomical and functional differences, there are hormonally induced psychological differences and differences created by society's influence and expectations. I even hazard the opinion that there are some vocations for which women may be better equipped then men, and—I hasten to add—vice versa.

6

SOME
FINAL
THOUGHTS

Now that we have completed our discussion of the control of the menstrual system and some of the social and psychological issues surrounding the menstrual cycle, there are a few matters left to consider in this final chapter.

the age of puberty . . . .

Recall that the initial start-up of the menstrual system (puberty) awaits the maturation of the hypothalamus and that higher brain centers may be involved also, since they have considerable influence on the hypothalamus and its control of the menstrual cycle. An interesting fact is that the age of puberty in Western Europe and the United States has decreased significantly in the last 100 years or so. In the middle of the nineteenth century, the average age for puberty was over 16; as indicated earlier, puberty now occurs at an average age of less than 13. The reasons for this lowered age of puberty are not known, but the brain may be at least partially responsible. At the beginning of Chapter 3 I stated that puberty requires that the ovaries become functional. The ovaries are dependent on the anterior pituitary which, in turn, requires input from the hypothalamus. But it is likely that the buck does not stop there but may be passed on still further to the higher sections of the brain, which, as was pointed out in Chapter 4, certainly do influence the hypothalamus.

Let's get back for a moment to the progressively earlier age of puberty over the past century. It has been suggested that nutrition may somehow be involved and that improved nutrition in the last 100 years is a factor in the earlier occurence of puberty. In fact, some researchers believe that puberty generally takes place when a girl's body weight reaches about 100 pounds. This is not an inflexible rule; the variation is great. Puberty taking place at a body weight of 85 or 90 pounds is not necessarily precocious, nor must it be considered to be delayed if it has not occurred at a body weight of 115 pounds or even more. The fact that there appears to be a correlation between nutrition and age of puberty does not necessarily mean that there is here a cause and effect relationship between the two. Indeed, we don't know exactly by what mechanisms improved nutrition and the earlier achievement of a particular body weight might serve as the cause of earlier function of the hypothalamus and, through it, of the menstrual system. Perhaps the next few years will provide some insight into this interesting issue.

sexuality and contraception . . . .

Puberty not only represents the time when menstrual cycles begin, it is also the time when sexual desires are intensified. Therein lies an obvious social problem. In the last century puberty, and the accompanying development of libido, occurred at an age when girls were more nearly adult—when they were emotionally more mature and thus better able to cope with developing sexuality. Today, with sex drive developing at an earlier age and marriage postponed to a later age, we have an expanded interval of sexual maturity without the sexual outlets traditionally considered acceptable by Western cultures. All this has had some influence on sex education and on ethical and legal views of sexual behavior; in addition, it is probably responsible, in part at least, for the so-called sexual revolution of the past decade or two, which has resulted in increased premarital sexual experience among young people in their teens and early twenties.

Sexual attitudes of females of all ages seem to have changed significantly, particularly in the last two or three decades. No longer are women willing to be regarded simply as sex objects or sexual outlets for men, and no longer must they suppress their own sexual needs. At long last women are regarded, both by themselves and by men, as equal partners, equal participants in an intimate and mutual experience. Not all men and women have yet come this far, but hopefully we are on the way.

These issues inevitably take us to a consideration of birth control. In spite of the ready availability of both information and methods for contraception, unwanted pregnancies still remain a problem.

Recently available data indicate that every year about 1 million girls in the United States between age 15 and 19 become pregnant; that represents almost 10 percent of the population of girls in that age range. About 30 percent of these pregnancies are terminated by abortion, which means approximately 300,000 teenage abortions per year. Obviously, abortion has become a significant method of birth control among teenagers. The same appears to be the case for older women as well.

Statistics such as these are undoubtedly a reflection of changing social standards regarding sexual activity among teenagers. However, regardless of one's moral, ethical, or religious convictions, it seems unnecessary to depend on abortion as a method of birth control when reasonably good methods of preventing pregnancy are available. In addition to being less private and more trouble than contraception, abortion is likely to entail greater physical and emotional trauma. This does not mean that abortions should not be available to girls or women who need or want them. Rather, abortion, as a regular method for birth control, is much, much less desirable than other available means that prevent pregnancy instead of terminating it.

The ideal means of contraception has not yet been developed. Each available method has its own particular advantages but possesses particular disadvantages as well. Some of the disadvantages may relate to safety, to limited acceptability, to inadequate effectiveness, and so on. Much has been written on the various features of the many methods presently used for preventing pregnancy. A review and evaluation of all or most of these methods are available elsewhere* and will not be repeated here. However, some contraceptive methods relate closely to the events of the menstrual cycle, and a brief description of those methods may contribute to your understanding of the cycle. Conversely, a knowledge of the menstrual cycle may increase your appreciation of the virtues as well as the shortcomings of some of the methods of preventing pregnancy. Therefore, the following discussion will emphasize those contraceptive methods that do relate to the cycle.

One method of preventing pregnancy is to prevent ovulation. Obviously, if an ovum is not released, there is nothing to be fertilized and pregnancy cannot take place. We now have hormonal treatments that do prevent ovulation—in other words, they produce anovulatory cycles. Such treatment involves the use of what is widely known as the (birth control) pill.

Both the ovarian hormones already discussed, es-

*Two sources of such information are: *Our Bodies, Our Selves* by the Boston Women's Health Book Collective (New York: Simon and Schuster, 1976); and *Woman's Choice* by Robert H. Glass and Nathan G. Kase (New York: Basic Books, 1970).

tradiol and progesterone, are represented in the pill. However, since neither estradiol nor progesterone is very effective when given by mouth, synthetic derivatives that are active when taken orally are present in all the pills now available. The effects of these synthetic compounds are almost exactly the same as those of the naturally occurring hormones, and the way they work to prevent ovulation is quite easy to understand.

In Chapter 3 it was mentioned that the contraceptive pill operates by taking advantage of feedback principles, and we can now examine that situation a little more closely. Recall that a prominent action of estradiol is its negative feedback on the anterior pituitary, and that FSH from the anterior pituitary is needed in order for the ovary to produce a mature follicle from which ovulation can occur. Estradiol, by its feedback action on the hypothalamus and anterior pituitary, tends to inhibit the release of gonadotropic hormones and, if given early in the cycle, will suppress the normal FSH output so that the ovary does not produce a mature follicle. Most contraceptive pills contain an orally effective estrogen and a progesterone substitute as well, and when they are taken daily for about 21 days out of every 28, the hypothalamus and anterior pituitary are inhibited, ovulation does not occur, and for that month the woman is sterile.

The ovary not only fails to release an ovum, but its hormone output is also markedly reduced. Remember that in the normal ovary the maturing follicle produces estradiol, and the corpus luteum produces both estradiol and progesterone. Without a maturing follicle, the estradiol output falls; in addition, a corpus luteum is not even formed unless ovulation has first occurred. However, the pills supply enough ovarian hormones to make up for the deficit in ovarian production. The pills must be taken every day except for about a seven-day interval each month. During the seven days when the pill is withdrawn, the endometrium, having lost its hormonal support, begins to break down, and a menstrual period occurs.*

Another cycle-related method of pregnancy prevention is commonly called the rhythm method, and it relies on the avoidance of sexual intercourse at and around the time of ovulation. This is not regarded as effective a method as use of the pill, but it may be more acceptable for some couples who, because of side effects or for religious reasons, prefer not to use the pill.

In order for a woman to be fertile, she must have an ovum in her oviduct ready to be fertilized; the fertile period in each menstrual cycle, then, is at or around the time of ovulation. Knowing when ovulation is to occur is consequently of importance to those couples who wish to rely wholly or partly on the rhythm method for conception control.

In Chapter 3 it was pointed out that, on the average, ovulation occurs about 14 days before the beginning of

*In recent years considerable concern has been generated about some of the side effects of birth control pills. Some pills have been withdrawn from use because of their alleged association with endometrial cancer. The most serious adverse effects of the contraceptive pills now in use involve apparent changes in the blood clotting system. These changes can lead to an increased frequency of strokes and heart attacks. High blood pressure also seems to be an occasional result of long-time use of the pill. Although the risks associated with use of contraceptive pills are not high, they do exist. Women over the age of 40 and those who are smokers run a greater risk than do younger women and nonsmokers.

the next menstrual period. This 14-day interval appears to be independent of the length of the cycle. (Take another look at that Figure 3.3 in order to refresh your memory.) Fourteen days is not a fixed and unchanging time; actually, the time of ovulation may vary from about 12 to 16 days before the next period. The 14-day interval is an average.

Unfortunately, unless the length of the menstrual cycle in a particular female is very constant, the time of the onset of the next period cannot be predicted with great accuracy—and one cannot then count the 12 to 16 days back to anticipate the time of ovulation. On the other hand, if accurate menstrual records are kept over a period of several months, and if the number of days between periods is constant within a day or two, then the time of ovulation can be anticipated. However, it must be emphasized that, even under those conditions, the accuracy of the prediction is open to some question.

Several means of determining the time of ovulation have been studied. Many women experience pain with ovulation (mittelschmerz), caused by slight ovarian bleeding. This is not always reliable, in that some women never seem to have mittelschmerz while others may experience it only occasionally.

Body temperature can also serve as a guide to the day of ovulation. If body temperature is carefully measured every morning before the woman gets up (rectal temperatures are sometimes recommended), a slight rise (amounting to about one-half degree Fahrenheit) occurs at ovulation time. Another means now sometimes used to detect the time of ovulation relies on changes in the character of the mucus that is commonly found in and around the cervix. In Chapter 3 we found that, at about the time of ovulation, cervical mucus, which is usually thick and viscous, becomes thin and watery, and a little may even leak out through the vagina. Some women are aware (or can teach themselves to be aware) of a sensation of increased vaginal wetness at this time.

It is believed that once ovulation has occurred, the released ovum remains fertilizable for only about 24 hours. However, it is important to remember that spermatozoa in the female reproductive system are believed to retain their ability to fertilize an ovum for about 48 hours, perhaps a little longer. Thus if intercourse takes place two days before ovulation, fertilization and pregnancy may result. Even if ovulation is detected as soon as it occurs, such information comes at least two days too late. Since the spermatozoa can retain their ability to fertilize for about two days, sexual intercourse must be avoided beginning more than two days *before* ovulation.

On the other hand, if intercourse occurs more than two days before ovulation or more than one day after ovulation, fertilization of the ovum is less likely. All this means that the normally cycling female has only about four days in each menstrual cycle when she is fertile—during the two days or so before ovulation, the day of ovulation, and the first day after ovulation. Those few days represent the most fertile period, and the remainder of the cycle is sometimes called the "safe period." The obvious question then is, How can the time of the fertile period be predicted? It cannot be done with precision, but a somewhat longer interval, which includes those four days, can be predicted with reasonable accuracy in many women.

Those couples who decide to make use of the rhythm method must do a little elementary arithmetic, after first keeping records on the length of the menstrual cycle for at least six months, preferably longer. If ovulation can occur as early as 16 days before a period, and the sperm cells can live for 2 days, then the potentially fertile period could begin 18 days before the next period. At the other end of the fertile period, since ovulation may occur as late as 12 days premenstrually and the ovum lives for an additional day, the fertile period may not end and the safe period begin until about 11 days before the next menstrual period.

In order to estimate the time in which the fertile period occurs, subtract 18 from the number of days of the shortest menstrual cycle and 11 days from the longest. The numbers obtained then give the time of the cycle when intercourse is most likely to result in pregnancy. Suppose, for example, that the menstrual cycles in a given female vary in length from 26 to 29 days. Subtracting 18 from 26 (shortest cycle) gives eight, and 11 from 29 (longest cycle) yields 18. This means that, in this example, days 8 to 18 of the menstrual cycle probably include the four-day fertile period and are thus "unsafe." Don't forget that day 1 of the cycle is defined as the first day of menstrual flow.

Many couples find that complete sexual abstinence for the duration of the "unsafe" period is difficult. Some have discovered that various sexual practices not involving actual entry of the female by the male may be quite satisfying during the fertile period, and such sexual contact of course involves no possibility of pregnancy.

Let me conclude this section on the rhythm method by emphasizing that the "safe period" is not absolutely safe. Pregnancy is much less apt to occur as a result of intercourse during the "safe period" than during the so-called fertile period; but one can't be certain because an unusually long or short cycle, an atypical time for ovulation, or an extended viability of ovum or spermatozoa could throw the calculations off. Several other methods of avoiding pregnancy are readily available—the condom, the diaphragm, the intrauterine device (IUD), and others.

One more very effective method of birth control is worth mentioning here—voluntary sterilization. This method is gaining in popularity and, according to some surveys, rivals the use of the pill. Among couples married 10 years or more, those who have chosen surgical sterilization appear to outnumber those for whom the wife still uses the pill. For young couples, particularly those who are newly married, the pill, not surprisingly, still predominates.

Voluntary sterilization involves a surgical procedure, and either the man or the woman can be sterilized. In recent years an increasing number of men have accepted sterilization as a means of birth control. The fear that such sterilization might decrease either their sexual appetite (libido) or their sexual ability served as a deterrent for some years, but most men now realize that such fears are unfounded. The surgical procedure for sterilizing a man is relatively simple, and it can be and frequently is performed in the physician's office. It merely involves cutting and tying the tiny tubes that conduct the spermatozoa from the two testicles to the penis. These tubes are known as the vas deferens, and the surgery is called a vasectomy.

Sterilization in a woman usually involves disrupting the two oviducts. Removal of the entire uterus (hysterectomy) has been used, but this operation represents major surgery. Unless the uterus needs to be removed for some other medical reason, the usual procedure is to tie the oviducts shut and to cut or cauterize them. This operation disrupts the canal through the oviduct and thus keeps the ovum and spermatozoa from getting together. Until a few years ago tying the oviducts also involved rather serious abdominal surgery, but new techniques have been developed that permit the operation to be carried out through a small hole in the abdomen not much bigger than a dime. The procedure in the male is still the simpler of the two.

In either the male or the female the sterility should be regarded as permanent, although some successful attempts have been made to reestablish the continuity of the vas deferens and the oviducts.

infertility . . . .

Most of us hear and read about the population explosion and the need for safe and effective methods of birth control. However, less well publicized is the problem of infertility, which appears to plague 10–15 percent of those couples who want children. Such apparent sterility can produce a considerable emotional burden on both the man and the woman who may desperately want children.

It is estimated that about 80 percent of couples will achieve pregnancy during the first year after abandoning contraceptives. Perhaps another 10 percent may succeed during the second year, and the remaining 10 percent may take much more time. If a couple tries for about a year and pregnancy does not occur, many physicians recommend that an investigation of the reasons for the apparent infertility should be started.

For a long, long time in many societies it was assumed (and still is in some) that when a woman failed to become pregnant, in spite of trying, the fault was always hers and never the male's. The fact is that approximately 40 percent of infertile marriages are the result of sterility in the male. Regardless of whether the defect resides in the male or the female, in a majority of cases appropriate treatment can and does result in a normal pregnancy.

Knowledge of the menstrual cycle can help in improving the likelihood of fertilization. Remember that in each menstrual cycle there are only about four days of fertility—in other words, only about four days when intercourse can result in pregnancy. So the couple may want to find those few days in order to have intercourse at a time when fertilization is most apt to take place. In this case the slight rise in body temperature associated with ovulation may be useful. This information comes a bit too late for those couples wanting to avoid pregnancy, but for those desiring pregnancy the temperature rise may occur just in time. It is estimated that the one-half degree rise occurs about 12 to 24 hours after ovulation, and spermatozoa deposited at that time may very likely find an ovum capable of being fertilized.

Appropriate timing of intercourse will not solve all

problems of apparent infertility. A number of physical defects may be present in either the male or the female. For example, inadequate numbers of spermatozoa produced and released by the male may be responsible for the failure to fertilize. In the female, anovulation, blocked Fallopian tubes, or cervical mucus hostile to spermatozoa are among the possible causes of sterility. In all these cases medical intervention frequently results in a correction of the defect and the desired pregnancy.

the "social regulatory factor". . . . .

One of the purposes of this chapter has been to present in some measure the interaction between the society we live in and the menstrual cycle. Issues such as birth control and infertility certainly relate to reproductive function in general and to the menstrual cycle somewhat more specifically; they also involve obvious social, cultural, and even religious issues. There is yet another social influence on the menstrual cycle, an intriguing influence I sometimes call the "social regulatory factor." This effect has been observed in both rats and humans.

   If a group of mature female rats obtained from a variety of different sources is placed in an isolated room in a single set of cages, the day of estrus for the individual rats will, as might be expected, be randomly distributed, with approximately the same number of rats going into estrus each day. But over a period of time, their estrous cycles seem to synchronize, so that after a few weeks perhaps 80 or 90 percent of them are having their estrous period on the same day. What is the factor that determines or controls this? No one knows.

Similar observations have been made on women. College women sharing dormitories have been found over a period of time to develop cycles in which their menstrual periods begin on or near the same day. In one particular study seven female lifeguards who worked together for an entire summer had menstrual periods scattered in time of onset at the beginning of summer, but by the end of the summer their periods were all beginning within a few days of one another.

Although no one knows by what mechanism the "social factor" operates, there is no doubt that somehow the brain must be involved. This is an intriguing but difficult subject to investigate. Again we may well ask, What is the stimulus, the input, to the menstrual system that changes the timing of the cycle to achieve synchrony? How did nature happen to develop this stimulus? Is it an unessential by-product of the organization of the system, or does it have some kind of biological value? These are legitimate questions, and further research may someday provide some answers.

changing times . . . .

In many ways the menstrual cycle has been a kind of burden that the female of our species has had to endure. In addition to withstanding some of the physical discomforts, the human female has frequently been subjected to menstrual-related discrimination, to a kind of contempt, and to isolation. She has been told that because of her cycle she is sometimes emotionally and intellectually unstable, that during her menstrual period she is unclean, that she must then be isolated and remain untouched, and that she may even be dangerous. These are some of the so-called menstrual taboos embraced by a number of primitive and some modern societies. In some societies it is believed that the menstruating woman is a danger to the entire community because she will cause the food supply to deteriorate. It has been said that the touch of a menstruating woman will kill plants, make wine or milk turn sour, and contaminate water. A very prevalent belief, still held in some societies, is that menstruating women are of special danger to men. Consequently, during their menstrual periods women have been banished to menstrual huts, have been prohibited from cooking for men, and have been shunned sexually. Some cultures disallow any sexual contact not only during the menstrual period but for several days afterward as well.

Biologists sometimes look at the male and female of any species as elaborate homes or containers for their germ cells, the spermatozoa and the ova. This point of view may have prompted one writer to declare, "A hen is only an egg's way of making another egg."* Certainly, nature has gone to great lengths to insure reproduction and the continuation of a species. It is deplorable, how-

*Samuel Butler, *Life and Habit*. Samuel Butler was a nineteenth-century English satirist and novelist who was also an amateur biologist. He was attracted to Darwin's theory of evolution, and *Life and Habit* is one of several books he wrote on this subject.

ever, that similar attitudes have pervaded many human societies, which regard women to be only "baby factories" and objects of pleasure for men.

Although many menstrual taboos and inequitable sexual attitudes are disappearing or are gone from advanced societies, I believe that we still have far to go. However, I am optimistic. Increasing knowledge may lead to wisdom and then to humane understanding. A friend of mine, an astute and yet optimistic historian,* has recently predicted the transcendence of humanity from *Homo sapiens*** to *Homo humanus*—a cultural evolution from a society of the wise to one of the humane.

Let it not be assumed that women must simply await decades of social evolution in order to gain their rightful role in society. In many ways women have become the authors of social change. Women have assumed increasing responsibilities, both in the family and in the community, and greater control over their destinies and over their own bodies.

This last thought brings me back to one more aspect of the control of the menstrual cycle and the very important role of the brain. We have seen how the higher centers of the brain have impact on the cycle and can even stop the cycle and prevent ovulation and the menstrual period. This great influence of the brain leads me to speculate about the possibility that the menstrual cycle might someday be brought under the control of conscious or deliberate will. Such an idea may seem preposterous, but in recent years evidence has been presented to

*L. S. Stavrianos, *The Promise of the Coming Dark Age* (San Francisco: W. H. Freeman and Company, 1976).
**Sapiens* is derived from a Latin word meaning wise or sensible.

show that such things as heart rate and blood pressure, earlier believed to be completely under automatic regulation, may be subject to some willful control. This kind of control is, in fact, under exploration for the treatment of certain types of heart disease. Imagine the importance of discovering a way of willfully preventing the LH surge and ovulation and thereby creating anovulatory cycles. Such an idea may sound like science fiction, but knowledge of the menstrual cycle has reached a point where this may be worth thinking about.

Willful control of the menstrual cycle may be a pipe dream today and may never become a reality, but a new look at the menstrual cycle and its significance is possible right now. Think back on what you have read in the preceding pages. Think of this amazing system that so beautifully integrates the brain, the anterior pituitary, the ovaries, and the uterus. All this belongs exclusively to the female, and it is all designed for the elemental and important function of periodically releasing a tiny egg and preparing a special place where a unique organism can be held, protected, and nourished while it develops into a new human being—perhaps *Homo humanus,* who, hopefully, will turn out to be both wise and humane.

GLOSSARY

Abdominal cavity   That part of the body containing the stomach, intestines, liver, and so on. It is sometimes called the belly.

ACTH   See *adrenocorticotropic hormone*.

Adenohypophysis   Another name for the anterior pituitary gland.

Adrenal cortex See *adrenal gland.*

Adrenal gland A complex endocrine gland located near the kidney. Each of the two adrenal glands may be divided into two parts: the cortex and the medulla. The cortex secretes cortisone; the medulla secretes epinephrine, better known as adrenalin.

Adrenalin See *adrenal gland.*

Adrenocorticotropic hormone (ACTH) One of the hormones produced by the anterior pituitary gland. It regulates the production of cortisone by the adrenal cortex.

Amenorrhea The abnormal absence of menstrual periods in a girl or woman of reproductive age.

Anovulatory cycle A menstrual cycle in which ovulation does not occur.

Anterior pituitary gland An endocrine gland located just below the brain that produces several hormones, including FSH, LH, and prolactin.

Catecholamines A group of compounds found in the brain, in other parts of the nervous system, and in the adrenal gland. They are involved in the transmission of signals in the brain and have many other effects as well. Adrenalin is a catecholamine.

Cerebrum The main part of the brain. It is divided into two halves, known as the cerebral hemispheres.

Cervix A word meaning neck. When applied to the uterus, cervix refers to the lower, narrow end that protrudes into the vagina.

Chorionic membrane An envelope that covers and protects the developing embryo and also produces hormones.

Circadian   A term applied to a biological cycle in which the events of the cycle occur every 24 hours. Similar to diurnal.

Coitus   Sexual intercourse.

Copulation   Sexual intercourse. The word *coitus* is more commonly used in referring to human sexual intercourse.

Corpus luteum   A yellowish glandular mass formed in the ovary from the remains of a follicle after it has ruptured and released its ovum. The corpus luteum produces estradiol and progesterone.

Corticotropin-releasing factor (CRF)   One of the hormones produced by the hypothalamus. It regulates ACTH release by the anterior pituitary.

Cortisone   One of the hormones produced by the adrenal cortex.

Diuretic   A substance that promotes the production of urine.

Diurnal   Recurring every day or having a daily cycle. Similar to circadian.

Dysmenorrhea   Painful menstrual period.

Edema   Accumulation of excessive fluid in the tissues of the body.

Endocrine glands   Those glands that secrete directly into the blood stream. The special substance so secreted is a hormone.

Endometrium   The lining of the inside of the uterus.

Estradiol   A hormone produced by the ovarian follicle, the corpus luteum, and the placenta. It is responsible for producing estrus in animals and for the development of secondary sex characteristics of the female.

Estrogen The general name given to substances that produce the same general effects as estradiol.

Estrogenic hormone See *estrogen*.

Estrogenic phase See *follicular phase*.

Estrous cycle The sex cycle in mammals other than the primates. This cycle includes a period of estrus characterized by sexual receptivity, or "heat."

Fallopian tube The tube into which the released ovum is transported and where fertilization by spermatozoa occurs. Also known as the oviduct.

Feedback The return of some of the output of a system as input to that system. Inhibitory, or negative, feedback tends to stabilize a system; positive feedback is reinforcing and leads to instability. In the menstrual system a positive feedback of estradiol leads to the so-called LH surge.

Follicle-stimulating hormone (FSH) One of hormones produced by the anterior pituitary gland; it induces growth of the ovarian follicle and the production of estradiol by that follicle.

Follicular phase A name sometimes assigned to that part of the menstrual cycle from the end of the menstrual period to the time of ovulation. This period is also known as the estrogenic phase of the menstrual cycle.

FSH See *follicle-stimulating hormone*.

Genital Pertaining to the reproductive organs.

Gonadotropic hormone A hormone that stimulates the activity of the gonads (ovaries or testes). FSH and LH from the anterior pituitary are gonadotropic hormones.

Gynecologist A physician who deals with reproductive function in women.

Hormone A chemical compound secreted directly into the blood by an endocrine gland and having an effect in some other part of the body.

Human chorionic gonadotropin (HCG) A gonadotropic hormone similar to pituitary LH but secreted by membranes surrounding the developing embryo.

Hypophysis Another name for the pituitary gland.

Hypothalamus A part of the brain located just above the pituitary gland and the source of several hormones that regulate the anterior pituitary gland.

Implantation The process by which the dividing fertilized ovum embeds in the endometrial layer of the uterus.

Ischemia Deficiency of blood in a structure of the body caused by constriction or obstruction of a blood vessel.

Lactogenic hormone See *prolactin*.

LH See *luteinizing hormone*.

Libido Sexual desire.

Luteal phase A name applied to that part of the menstrual cycle that extends from just after ovulation to the beginning of the next menstrual period. The luteal phase is also known as the progestational phase.

Luteinization The process whereby the cells remaining in a ruptured follicle after ovulation multiply and enlarge to form a corpus luteum.

Luteinizing hormone (LH) One of the hormones secreted by the anterior pituitary; it induces ovulation and corpus luteum formation.

Luteinizing-releasing factor (LRF) One of the hypothalamic hormones. It controls the release of both LH and FSH by the anterior pituitary gland.

Mammary glands   The breasts or, more specifically, the milk-producing glands of the breasts.

Menopause   The final or irreversible cessation of menstrual cycles in the human female.

Menstrual cycle   The processes that result in a periodic release of an ovum and a loss of blood and tissue from the uterus.

Menstrual period   That part of the menstrual cycle when blood is discharged from the vagina.

Mittelschmerz   Lower abdominal pain associated with ovulation.

Mucosa (mucous membrane)   The lining of a body cavity. The endometrium is an example of a mucosa.

Myometrium   The muscles of the uterus.

Neurohypophysis   The posterior pituitary gland.

Nidation   See *implantation.*

Oncologist   A tumor or cancer specialist.

Orgasm   The climax of sexual excitement marking the culmination of the sex act and following which release of sexual tension occurs.

Ovarian cycle   The periodic changes in the structure and function of the ovary related to the menstrual cycle.

Ovary   The sex gland of the female; the source of egg cells (ova) and of estradiol and progesterone.

Oviduct   See *Fallopian tube.*

Ovulation   The release of an ovum from an ovarian follicle.

Ovum   The reproductive cell of the female; an egg cell.

Oxytocin   One of the hormones released by the posterior pituitary gland. It causes contraction of uterine muscles and of contractile elements in the breast.

Parturition The process of giving birth.

Pituitary gland An endocrine gland situated just below the part of the brain known as the hypothalamus. Among the several hormones produced by the pituitary are follicle-stimulating hormone (FSH), luteinizing hormone (LH), and prolactin (PRL) from the anterior pituitary; and oxytocin from the posterior pituitary.

Placenta The organ attached to the uterus of the pregnant woman that (by way of the umbilical cord) attaches the developing fetus to the mother. It is discharged from the uterus shortly after the infant is born and is commonly called the afterbirth.

Premenstrual syndrome The collection of symptoms that rather commonly occur in the few days before a menstrual period.

PRL See *prolactin*.

Progestational phase See *luteal phase*.

Progesterone One of the hormones produced by the ovary. Its source is the corpus luteum, and thus it is found in the blood in substantial quantities only during the second half (luteal phase) of the menstrual cycle.

Prolactin (PRL) A hormone produced by the anterior pituitary and also known as lactogenic hormone. It is responsible for inducing milk production by the mammary glands.

Prolactin-inhibiting factor (PIF) One of the hormones produced by the hypothalamus. It inhibits the release of prolactin by the anterior pituitary gland.

Pseudocyesis False pregnancy, sometimes called phantom pregnancy.

Puberty The period when sexual maturation occurs.

Pubescent Arriving at the age of puberty.

Receptor sites A chemical complex in or on a cell that has the ability to bind and interact with substances such as hormones.

Somatostatin A hormone produced by the hypothalamus that inhibits growth hormone release by the anterior pituitary.

Somatotropin A hormone produced by the anterior pituitary; more commonly known as growth hormone.

Spermatozoa The male germ cells, produced by the testes and required for fertilizing an ovum.

Sperm cells See *spermatozoa*.

Testicle See *testis*.

Testis The male sex gland, which produces spermatozoa as well as a male hormone.

Testosterone The hormone produced by the testis. It is responsible for the development of male secondary sex characteristics.

Thyrotropin A hormone produced by the anterior pituiary and required for the normal function of the thyroid gland. It is also known as thyroid-stimulating hormone (TSH).

Thyrotropin-releasing factor (TRF) A hormone produced by the hypothalamus that promotes release of thyrotropin (thyroid-stimulating hormone) by the anterior pituitary.

Uterus The organ in the female where the embryo is maintained previous to birth. Also known as the womb.

Vagina   The canal extending from the cervix of the uterus to the outside.

Vasomotor   Affecting the diameter, or the caliber, of a blood vessel.

INDEX

Amenorrhea, 73–86
  anorexia nervosa, 83–84
  college amenorrhea, 83
  false pregnancy, 80–83
  lactation, 78
  menopause, 78–79
  pregnancy, 74–77
Anovulatory cycles, 70–71

Birth control. See
  Contraception
Brain, 28–39, 80–81, 88. See
  also Hypothalamus
Breasts, 24–26, 66–67, 70,
  99. See also Milk
  production

Cervix and cervical mucus,
  15–16, 67–68, 112–113
Contraception, 108–116
  the pill, 109–111
  rhythm method, 111–115
  sterilization, 115–116
Corpus luteum. See Ovaries

Dysmenorrhea, 93–96

Endometrium, 17, 57–58,
  64–66, 75, 77, 111
Estradiol, 19–20, 22, 25, 34,
  37, 49, 54, 57–59,
  60–62, 64–65, 68, 77,
  98, 110. See also Ovaries

Estrogen. *See* Estradiol

Estrus period, 6–7

Fallopian tube (oviduct), 27, 63–64, 75

False pregnancy. *See* Psychological factors in the cycle

Feedback control, 45, 48–50, 60

Fertilization, 75

Follicle-stimulating hormone (FSH). *See* Pituitary gland

Gonadotropic hormones. *See* pituitary gland

Hot flushes, 97

Hypothalamic-Hypophyseal Portal system (HHPS), 30–31

Hypothalamus, 29–33, 35, 43–44. *See also* Luteinizing hormone-releasing factor (LRF)

Infertility, 116–118

Lactation. *See* Milk production

Length of menstrual cycles. *See* Menstrual system

LH surge, 62, 70

Libido, 7, 89

Light, effect of on cycle, 3–4, 8–9

Luteinizing hormone. *See* Pituitary gland

Luteinizing hormone-releasing factor (LRF), 32, 33, 37, 45, 48, 49–50, 54, 58–63, 79

Menopause, 73, 78–79, 96–102

Menstrual period, 18, 51, 54–55, 68–70, 75, 111

Menstrual system, 41–50
communication pathways, 11–14
defects, 85–86
estrogenic (follicular) phase, 51–59, 71
progestational (luteal) phase, 51–54, 64–68
timing (cycle duration), 9, 50–54, 74

Milk production, 34–36, 66–67, 78

Mittelschmerz, 21–22, 63, 112

Myometrium, 17–18, 57, 65, 94–95

Ovarian follicles. *See* Ovaries

Ovaries, 18–22, 55–56, 99
  corpus luteum, 22, 27–28,
    65, 68, 78
  follicles, 19–20, 27, 55, 75,
    110
Ovulation, 4, 5–6, 19–22,
  51–54, 55–57, 59–64,
  111–112
Oxytocin, 35–36

Pituitary gland, 22–28. See
  also Prolactin; Oxytocin
  follicle-stimulating
    hormone (FSH), 27,
    36–37, 50, 54–55,
    79, 110
  luteinizing hormone (LH),
    27–28, 36–37, 50, 55,
    58–59, 62–63, 64, 70,
    77
Placenta, 34, 77
Pregnancy, 74–77
Premenstrual discomfort,
  90–93
Progesterone, 22, 25, 34, 37,
  54, 60–62, 65, 68, 77.
  See also Ovaries

Prolactin, 24–25, 26, 33–34,
  78
Psychological factors in the
  cycle, 79–84, 102–104
  anorexia nervosa, 83–84
  college amenorrhea, 83
  false pregnancy, 80–83
  in menopausal symptoms,
    97–100
  in premenstrual
    symptoms, 92–93,
    95–96
Puberty, 43–45, 73, 106–107

Safe period, 113–115. See
  also Contraception
Seasonal reproduction, 2–4
Social factors in the cycle,
  99, 101, 118–122

Timing of cycle. See
  Menstrual system

Uterus, 15–18. See also
  Endometrium;
  Myometrium

Vagina, 15–16, 100